GOING · GREEN

GOING · GREEN

People, politics and the environment in South Africa

Edited by
Jacklyn Cock
&
Eddie Koch

1991
OXFORD UNIVERSITY PRESS
CAPE TOWN

OXFORD UNIVERSITY PRESS
Walton Street, Oxford OX2 6DP, United Kingdom

Oxford New York Toronto
Delhi Bombay Calcutta Madras Karachi
Petaling Jaya Singapore Hong Kong Tokyo
Nairobi Dar es Salaam Cape Town
Melbourne Auckland
and associated companies in
Berlin Ibadan

ISBN 0 19 570582 3

Editing: Helen Laurenson
Design: Mara Singer and Welma Odendaal
Profile interviews: Pat Schwartz
Page make-up and layout: Welma Odendaal
Artwork: Nick Curwell and Mara Singer
Cover: Sue Williamson and Mara Singer
Proofreading: Helen Moffett

Published by Oxford University Press Southern Africa, Harrington House, Barrack Street, Cape Town 8001, South Africa
Set in 10 on 12 point Caslon
Reproduction Theiner Typesetting and Clyson Printers
Printed and bound by Belmor Printers

We would like to thank Glenda Younge of Oxford, who initiated this project and fired us with her enthusiam, and Helen Laurenson, our editor, who has been a delight to work with.
The Editors

The editors and publisher gratefully acknowledge the photographers whose work was used in this book: their credits appear next to the relevant photographs. In some cases we have been unable to trace copyright holders and would be grateful for any information that would enable us to do so.

CONTENTS

PREFACE

LIST OF CONTRIBUTORS

1 **GOING GREEN AT THE GRASSROOTS** ... 1
The environment as a political issue
◆ JACKLYN COCK ◆

PROFILE — CHRIS ALBERTYN ... 18

2 **RAINBOW ALLIANCES** ... 20
Community struggles around ecological problems
◆ EDDIE KOCH ◆

3 **RISKING THEIR LIVES IN IGNORANCE** ... 33
The story of an asbestos-polluted community
◆ MARIANNE FELIX ◆

PROFILE — TSHEPO KHUMBANE ... 44

4 **THE GHETTO AND THE GREENBELT** ... 46
The environmental crisis in the urban areas
◆ LESLEY LAWSON ◆

PROFILE — JAPHTA LEKGHETO ... 64

5 **VICTIMS OR VILLAINS?** ... 66
Overpopulation and environmental degradation
◆ BARBARA KLUGMAN ◆

6 **REDS AND GREENS** ... 78
Labour and the environment
◆ ROD CROMPTON ◆ ALEC ERWIN ◆

PROFILE — HUMPHREY NDABA ... 92

7 **THE IMBALANCE OF POWER** ... 94
Energy and the environment
◆ MARK GANDAR ◆

PROFILE — MARK GANDAR ... 110

8 **FLOWERS IN THE DESERT** ... 112
Community struggles in Namaqualand
◆ DAVID FIG ◆

9 WASTING WATER ... 129
Squandering a precious resource
• HENK COETZEE • DAVID COOPER •

10 THE INSANE EXPERIMENT ... 139
Tampering with the atmosphere
•JAMES CLARKE •

PROFILE — JENNY MUFFORD ... 158

11 WASTING AWAY ... 160
South Africa and the global waste problem
• PETER LUKEY • CHRIS ALBERTYN • HENK COETZEE •

PROFILE — BEV GEACH ... 174

12 FROM SOIL EROSION TO SUSTAINABILITY ... 176
Land use in South Africa
• DAVID COOPER •

13 THE OCEANS ... 193
Our common heritage
• FRANCIS MANUEL • JAN GLAZEWSKI •

PROFILE — NAN RICE ... 208

14 THE CRUCIAL LINK ... 210
Conservation and development
• MARGARET JACOBSOHN •

15 ANIMALS VERSUS PEOPLE ... 223
The Tembe Elephant Park
• THE ASSOCIATION FOR RURAL ADVANCEMENT •

PROFILE — IAN PLAYER ... 228

16 BIODIVERSITY: THE BASIS OF LIFE ... 230
South Africa's endangered species
• JOHN LEDGER •

PROFILE — AILEEN TAYLOR ... 242

17 THE TEETH NEED SHARPENING ... 244
Law and environmental protection
• JON WHITE •

SELECTED REFERENCES ... 252

INDEX ... 257

PREFACE

This book is being published at an exciting time in South Africa. A powerful environmental movement is beginning to emerge. This movement is evident in different communities and social groups who are mobilizing around a wide range of issues, ranging from asbestos dumps to water and air pollution, from the ivory trade to soil erosion, from lead-free petrol to herbicides and from nuclear energy to toxic waste. Together we represent a great wave of South Africans who are concerned for the environment, and thus for the future and for our children.

This book also appears at a critical time in South Africa's history. We are engaged in creating an entirely new and different society. Nadine Gordimer has called this 'one of the most extraordinary events in world history — the complete reversal of everything that, for centuries, has ordered the lives of all our people'.

The creation of a just and peaceful society will not be easy. Clearly one of the priorities of a new democratic government must be to eliminate poverty and improve the quality of life for the millions of our people who have been subjected to centuries of discrimination and denial. An important example of this will be to provide electricity to the 60 per cent of South Africans who do not have it at present. The nature of a cost-effective, sustainable and environmentally-friendly energy policy must be the topic of widespread debate in the new South Africa.

Such debates — in which the notion of 'sustainable development' must be central — need to be anchored in a mass environmental awareness. We are moving into a new order in which ordinary people will hopefully have the power to shape their future, including the formulation of sound environmental policy. This book is intended to contribute towards the creation of such an awareness. Our contributors are a diverse group of people, but this is our common aim.

South Africa is indeed 'a world in one country'. We are one of the world's richest areas of biodiversity. But with our mix of First World environmental problems such as acid rain, and Third World environmental problems such as soil erosion, we are a microcosm of the environmental challenges facing the planet. We are also a microcosm of the inequalities behind the division into 'First World' and 'Third World' — a situation in which the richest 20 per cent of the world consumes 80 per cent of the world's resources.

A global perspective is necessary if we are to address these inequalities. There are real dangers that the 'greening' of western countries will lead to the exporting of the environmental crisis and its intensification in the Third World. In recent years there has been a very dramatic increase in the export of toxic chemical and industrial waste.

The slogan, 'Think global, act local,' is a key theme in our struggle. As isolated individuals it is easy to feel overwhelmed by the scale of environmental abuse. But if we act together we can deepen our understanding and develop our collective strength. We therefore urge our readers to inform themselves and become active in the range of organizations now addressing environmental issues. The environmental movement in South Africa urgently needs *your* time and *your* voice.

— *The Editors*

LIST OF CONTRIBUTORS

CHRIS ALBERTYN is a student counsellor at the University of Natal, Pietermaritzburg, and a founder member of Earthlife Africa. He has a special interest in social ecology.

JAMES CLARKE is the Assistant Editor of *The Star* and a veteran writer on environmental matters.

HENK COETZEE is a geologist and a founder member of Earthlife Africa. He is also a researcher for the Environmental Monitoring Group.

JACKLYN COCK is a sociologist at the University of the Witwatersrand. She is the author of *Maids and Madams* and co-editor with Laurie Nathan of *War and Society: The Militarization of South Africa*.

DAVID COOPER is an agriculturalist, a founder member of the Environmental and Development Agency (EDA), and coordinator of the Environmental Monitoring Group.

ROD CROMPTON is General Secretary of the Chemical Workers' Industrial Union (CWIU), a union that has played a leading role in involving organized labour in environmental campaigns.

ALEC ERWIN is the educational officer of the National Union of Metal Workers of South Africa (NUMSA), a veteran trade unionist and leading economic theorist.

MARIANNE FELIX is a medical doctor with a special interest in occupational health. She has done pioneering research in the field of environmental health.

DAVID FIG is a political scientist with interests in labour, international relations and nuclear energy. He serves on the board of *New Ground*, the environmental journal produced by the Environmental and Development Agency (EDA).

MARK GANDAR is a nuclear physicist, founder of the Society Against Nuclear Energy (SANE) and at present a consultant in the use of sustainable energy.

JAN GLAZEWSKI is an environmental lawyer at the University of Cape Town who has published widely on constitutional law and the environment.

MARGARET JACOBSOHN is a social anthropologist who is currently based in northern Namibia. At present she is director of community development for a worldwide nature and endangered wildlife fund. She is the author of *Himba: Nomads of Namibia.*

EDDIE KOCH is the Environment Editor on *The Weekly Mail* and co-author with Henk Coetzee and David Cooper of *Water, Waste and Wildlife: The Politics of Ecology.*

BARBARA KLUGMAN is a social anthropologist at the Centre for Health Policy Studies, University of the Witwatersrand. She is currently working in the area of women and health and is one of the authors of *Vukani Makhosikazi: South African Women Speak.*

LESLEY LAWSON is a writer, photographer and film-maker. She is the author of *Working Women* and is currently involved in the production of an environmental film series.

JOHN LEDGER is the Director of the Endangered Wildlife Trust and a conservationist of long standing. He has a special interest in maintaining biodiversity and achieving community involvement in conservation issues.

PETER LUKEY is a civil engineer, a founder member of Earthlife Africa and a veteran of the green movement in South Africa.

JON WHITE is an attorney in Pietermaritzburg. He is active in Earthlife Africa and has a special interest in law and the environment.

1

GOING GREEN AT THE GRASSROOTS

The environment as a political issue

◆ JACKLYN COCK ◆

South Africa, with its mix of First World environmental problems such as acid rain, and Third World environmental problems such as soil erosion, is a microcosm of the environmental challenges facing the planet. These challenges are slowly coming onto the agenda of the liberation struggle in South Africa.

The authoritarian conservation perspective

Until very recently the dominant understanding of environmental issues in South Africa was an authoritarian conservation perspective. This focused exclusively on the preservation of wilderness areas and particular species of plants and animals. Within this perspective 'overpopulation' was often identified as the main environmental problem. It was people who were perceived to be responsible for destroying trees and creating waste.

Only very recently has an alternative progressive perspective begun to emerge. This perspective views environmental issues as deeply political in the sense that they are embedded in access to power and resources in society. It is critical of the victim-blaming approach of the overpopulation perspective and insists on the need for development to overcome poverty. It draws on the ideology of 'green politics' to emphasize the importance of linking the struggle against social injustice and the exploitation of people with the struggle against the abuse of the environment.

There are similarities between violence against people and violence against the environment

The legacy of the authoritarian conservation perspective is that many South Africans view environmental issues with suspicion. Farieda Khan has pointed to 'the negative environmental perceptions and attitudes of many black people, ranging from apathy to hostility' (Kahn, 1990:37). In the past, conservation projects often disregarded human rights and dignity. The establishment of many game reserves meant social dislocation and distress for local people. As a rural worker has stated, 'If conservation means losing water rights, losing grazing and arable land and being dumped in a resettlement area without even the most rudimentary infrastructure and services, as was the case when the Tembe Elephant Park (near Kosi Bay) was declared in 1983, this can only promote a vigorous anti-conservation ideology among the rural communities of South Africa' (Richard Clasey, a rural fieldworker quoted in *The Weekly Mail*, 6 October 1989).

These authoritarian practices are not safely buried in the distant past. Even today environmental campaigns are often directed at a restricted middle-class audience. For example, the Autumn 1990 issue of the recently launched Cape Coast environmental magazine, *Coastline*, contains an article which starts, 'It is almost a truism to say that if one lives inland one aims to get to the coast for a holiday at least once a year and later to retire there'. There are numerous instances of this kind of middle-class bias and insensitivity in the literature on environmental problems. Furthermore, environmental issues are not always directed at a non-racial audience. For example, the Transvaal Provincial Administration's Earth Day 1990 festivities were held at the 'whites only' Overvaal resort at Swadini in the eastern Transvaal. One of the organizers of the event, Mrs Hester Theron, information officer at the Transvaal's Department of Nature Conservation at Swadini, said Overvaal's race policy had 'unfortunately been overlooked' when the event was planned (*The Star*, 20 April 1990).

The authoritarian conservation perspective tends to focus on the conservation of cuddly or spectacular creatures like the blue whale or the tiger or the giant panda bear. There are conservationists who sometimes sound a little misanthropic and appear to be more concerned with animals than people. Infant mortality rates in the 'homelands' do not seem to have the same fashionable appeal as the conservation of the black rhino. The result is that many South Africans view environmental issues as white middle-class concerns.

PICTURE PROVIDED BY TOWN CLERK OF CROSSROADS

People in the environment

However, environmental issues do have a relevance to all people. Environmental issues do have 'the potential to build alliances across the divides of class and race' as Koch and Hartford have argued (*The Weekly Mail*, 6 October 1989) — they cite the indiscriminate use of pesticides as an example. However, there is no smooth and easy convergence of class and race interests around this issue. In the first place, the pesticide industry is a source of enormous profit to a minority. Secondly, the vegetable farmers in Natal who have suffered from the indiscriminate use of pesticides such as the Agent Orange-type herbicide, are able to mobilize public opinion, even though they failed to win a recent Supreme Court application to prohibit the manufacture and sale of all hormone herbicides in South Africa. Middle-class consumers have access to knowledge of

South Africa is a microcosm of the environmental challenges facing the planet

ABOVE
Squatter housing on the Cape Flats in Cape Town

the link between pesticides and cancer and have the purchasing power to shift to organically-grown produce bought at expensive health food shops. The real victims are the poor, who do not have either this knowledge or this option. In their ranks are, as Koch and Hartford say, the agricultural workers who spray the pesticides as well as the factory workers who manufacture them.

Toxic waste is another issue that affects us all. However, the people most affected are the workers directly handling these hazardous materials. There are reports that some workers at Thor Chemicals, a multinational company which imports highly toxic mercury waste into South Africa, are suffering from a severe nervous disorder induced by mercury poisoning. In a nearby river, mercury concentrations have been recorded at hundreds of times the level at which the USA's Environmental Protection Agency (EPA) declares materials toxic. It is the poor who suffer most from such pollution of their water supplies.

It is also the poor confined to townships such as Soweto who suffer most from air pollution — not only from power plant emissions but also from coal stoves. According to ESKOM, pollution levels in Soweto are 2,5 times higher than anywhere else in South Africa. Medical studies show that Soweto's children suffer from more asthma and chest colds, and take longer to recover from respiratory diseases, than do youngsters elsewhere in the country.

The state and the environment

The 'progressive developmental perspective' rooted in the ideology of 'green politics' (Capra and Spretnak, 1984; Bahro, 1984; Porritt, 1984; Tobias, 1989) links the struggle against such poverty and social injustice with the struggle against the abuse of the environment. In South Africa this connection means that environmental issues are deeply and fundamentally political. Ironically, a Wildlife Society survey of candidates in the last tricameral election in 1989 found that the majority of those who replied believed South Africa has no real environmental problems. Of the candidates who replied 37 per cent rated 'visiting the Kruger National park' of greater conservation importance than 'improving the quality of life'.

The leader of the Conservative Party, Andries Treurnicht, believes ecological issues are 'not so important that they will be basic to our policy' (*IDASA Newsletter*, October 1990). Recently the Ecology party was formed with the aim of operating outside 'politics' to promote environmental awareness. The Democratic Party spokesman on the environment, Rupert Lorimer, has commented that this is not really feasible because 'ecology is not the only issue' (*The Star*, 28 October 1989). In fact, ecology is *the* issue, but it is deeply embedded in a mass of other issues concerning the distribution of power and resources in South Africa. Green politics here has to be firmly anchored in the needs of the majority of our people. For the majority of our people environmental issues mean no clean water, no electricity and no proper sanitation. These are the issues which need to be addressed as a matter of urgency. This can be done only by a democratic government which is accountable to the people and which prioritizes their interests.

Instead we in South Africa have a deeply authoritarian, repressive and undemocratic government. The contempt our ministers have for ordinary people is illustrated by the response of the ex-Minister of Environmental Affairs, Mr Gert Kotze, to *The Star's* petition to save St Lucia, which had then passed the 200 000 mark. The minister was unimpressed. He told a Cape audience that 'most of these people did not know what they were signing', that such people caused development to be delayed, and 'as we all know, time is money'. He also said that half the signatures were from children (*The Star*, 9 November 1989). In the Minister's opinion, people who sup-

ABOVE
Workers dispose of waste products illegally at a dump outside Johannesburg

JUSTIN SHOLK

port 'green' movements are 'fanatics who do not listen to reason' (*The Saturday Star*, 11 November 1989).

Such arguments have often been used against environmentalists, even in other less repressive societies. Rachel Carson's book *The Silent Spring*, which was the catalyst for the American environmental movement, was termed to be part of a communist plot to undermine the American economy. Carson was dismissed as a 'spinster who should have no concern for genetics' (Hynes, 1989).

Protecting the environment

In South Africa, although the new Environmental Conservation Act has been hailed as an improvement, we desperately need more effective state regulation, control and protection of the environment. Our natural resources are exceptionally diverse, with over 240 species of mammals, 887 different species of birds and over 20 000 species of flowering plants. These diverse resources are not adequately protected.

Clearly we need something like the American Environmental Protection Agency. This was founded on two principles:

- People and nature have rights to exist unendangered by pollution and should be enabled to do so by regulatory protection.
- This can be done only by means of a framework which is set up to control polluting industries whose primary imperatives have been to compete for markets and to increase profits.

However, the problems of enforcement experienced by the EPA are well known. Environmental laws are effective as safeguards only if they are creatively and vigorously enforced. This requires environmentally committed and ethical persons who are not bribable and cannot be intimidated. In South Africa such persons are in short supply. The cycad scandal has pointed to official corruption. Around 700 of these protected, rare, South African plants suddenly arrived at a private home in Madeira without the proper export and import licenses. It is frequently said that South Africa is a conduit and clearing house for ivory and rhino horn poached in countries as far away as Tanzania and Zaïre. And of course here the problem is not only about official passivity or corruption, but also about the drive for profit.

Capital and the environment

The recent book by Brian Huntley, Roy Siegfried and Clem Sunter, *South African Environments into the Twenty-First Century*, is structured around the notion of choice — clarifying certain choices the reader has to make between different scenarios. However, there are two highly controversial assumptions in the book which are not opened up to choice; instead they are treated as givens. The first is that nuclear energy is necessary and desirable. The second is that the free enterprise system is the only viable one. These are both assumptions which need to be exposed and subjected to critical scrutiny and debate.

There are many serious people who believe that nuclear energy is too expensive and unsafe to be a viable alternative energy source for the future. (See Chapter 7.) They say that the human factor involved in all stages of nuclear technology, military and non-military, makes accidents unavoidable. Many nuclear accidents have already happened, releasing highly poisonous radioactive materials into the environment. It is now acknowledged that the accident at Chernobyl released many times more deadly radioactivity into the atmosphere than did the explosion of the nuclear bomb at Hiroshima. A debate about nuclear power in South Africa is an urgent necessity given the secrecy surrounding the nuclear industry and the missionary zeal with which ESKOM discusses its plans for the expansion of its nuclear programme (Semark, 1990).

The total explosive power of the world stock of nuclear weapons is about equal to one million Hiroshima bombs

Another major problem of nuclear power is the disposal of nuclear waste. Each reactor annually produces tons of radioactive waste that remains toxic for thousands of years. Nuclear waste from Koeberg includes at least 200 kilograms of plutonium a year. Plutonium is so toxic that 5 kilograms is enough to kill every man, woman and child on this earth (Henk Coetzee of Earthlife Africa, quoted in *The Weekly Mail*, 4 May 1990). Plutonium remains poisonous for at least 500 000 years: no human technology can create safe containers for such an enormous time span. (See Chapter 11.)

Huntley, Siegfried and Sunter's other unexamined assumption relates to the free enterprise system. They assert that 'the greens have gone wrong. The greens contradict themselves by demanding socialism and a clean environment at the same time. Socialism leads inevitably to the malfunctioning of the economy, which means that no money is available for conservation ... Man is a born opportunist. By denying that self-interest is a valid principle, socialists are suppressing people's innate ability to think in an efficient manner ... communism is an unnatural philosophy inflicted on mankind' (Huntley et al, 1989:14-15).

This is a very contentious argument. There are many serious people who argue that one simply cannot cure today's problems with the means that have produced them. Many people believe that it is capitalism, an economic system obsessed with growth and expansion, that has created many of our environmental problems. Jonathan Porrit, the Director of the Friends of the Earth, has argued that capitalism means ecological disaster. 'Capitalism can indeed survive only through permanent expansion — which in turn means the accelerating contraction of our life-support systems' (Porrit, 1984:225). 'Capitalism as we know it simply cannot provide the preconditions for an ecologically sane, humane economy' (Porrit, 1984:48).

In South Africa it is capitalists who are climbing on 'the green bandwagon' to promote consumption. It is capitalists who, in the name of 'growth' and 'development' are destroying our coastline and restricting its use to those who can afford to buy property in their costly marina developments. However, in the global context it is not only capitalists who are destroying the finite resources of our finite planet in their quest for profit. The environments of Eastern Europe and the Soviet Union have suffered greatly from the intensive industrialization of the post-war period. The drive to meet production quotas has proved as environmentally disastrous as the profit drive.

Infant mortality rates in the 'homelands' do not seem to have the same fashionable appeal as the conservation of the black rhino

Doing violence to the environment

It is widely agreed that to achieve the aims of green politics and end the exploitation of both people and the environment, change is required. But what is hotly contested is the nature of the change. Bahro argues for a dismantling of industrial society spearheaded by the 'eco-pacifist movement' (Bahro, 1984:229). The more general view is that the problems can be put right by a few adjustments to the system — less pollution, less destruction of vital resources and more environmentally-conscious consumption. The problems 'are to be wafted away through the mechanism of the very force that created them, and without anybody having to make any very substantial changes ... There is no suggestion anywhere that we should stop producing vast amounts of unnecessary electricity with coal plants that create acid rain; no thought that we should change our absurd dependence on the private automobile even though it is bound to produce toxic emissions no matter what miracle fuels we come up with; no hint that we should eliminate the industries that routinely produce toxic chemicals' (Kirkpatrick Sale, *Resurgence*, September/October 1989).

NAMAQUANUUS

Sale is one of a number of environmentalists who are urging middle-class people to change their lifestyles, to reduce consumption, to move away from a consumer to a conserver economy. In the same way that feminism or gender equality is not compatible with male chivalry and protection, it is clear that a beautiful, unspoilt environment is not compatible with the present high levels of consumption in the developed world. We cannot have it both ways. As Schumacher has stated, 'We must live simply that others may simply live'.

Schumacher and Sale have advocated a 'voluntary simplicity'. They point to a number of reasons why a simplification of life in the First World is important:

- We are running out of crucial non-renewable resources. For example, we are exhausting the supply of cheaply available petroleum and natural gas.
- We are polluting ourselves steadily into oblivion with massive discharges of wastes from industrial production.
- Each year 14 million children under the age of five in the Third World die from the effects of malnutrition and poor health care (Timberlake and Thomas, 1990). They do so partly because of the massive military expenditure which absorbs so much of the world's resources.

ABOVE

Representatives of various organizations in Namaqualand rally around the issue of forced removals at a National Union of Mineworkers congress in Bergsig in May 1989

One of the greatest threat to our environment is war and the mobilization of resources for war. War has devastating effects on the environment. This is particularly clear in the southern African region, where the economies of Mozambique, Angola and Namibia have suffered from prolonged war waged by the SADF or its surrogate forces. The threat lies not only in the awesome destructive capacity of the weapons mankind has developed for war, but in the resources invested in this process. Worldwide military expenditure now amounts to more than 900 billion dollars per year. There are more than 50 000 nuclear warheads in the world. The total explosive power of the world stock of nuclear weapons is about equal to one million Hiroshima bombs. There are at least 50 million people throughout the world who are either directly or indirectly engaged in military activities. In South Africa the Department of Defence received just under 20 per cent of the total budget for 1990/1, while the Department of Environmental Affairs received one third of 1 per cent.

Clearly there has to be a massive shift of resources away from military projects into developmental activities such as energy efficiency, soil conservation and tree planting. But the connection between war and environmental destruction is deeper than this.

There are similarities between violence against people and violence against the environment. Both are about maintaining power and maximizing profit. Both kinds of violence involve similar technology: for example, there are many similarities between nuclear energy and nuclear weapons. They involve the same raw materials and similar technologies. Another example relates to chemical weapons and pesticides. In *The Silent Spring* Rachel Carson talked about how the war on people and the war on nature often employed the same weapons. Nerve gases developed for World War II were used as pesticides in agriculture after the war. Likewise, herbicides developed for agriculture before the Vietnam War were used as defoliants in that war, and by South Africa in Angola. Carson's biographer, Patricia Hynes, writes, 'The destruction of people and nature with chemical poisons constitutes the same failure to solve problems other than by force.' Carson's central warning was that the methods employed for insect control may be such that 'they will destroy us along with the insects'.

In 1985 the 'Hiroshima of the chemical industry' occurred when an accident at a pesticide factory in Bhopal, India, caused the death of at least 3 000 people and injured 200 000. Clearly violence against nature and violence against people are connected.

Finally, both kinds of violence involve a similar set of attitudes which legitimate killing and violence as a solution to conflict; both value domination, conquest and control; both involve a disrespect for human and other forms of life.

This 'disrespect' is evident in the plans to mine the dunes at St Lucia for titanium. Besides being highly profitable to the company involved, titanium is a mineral that is used in missile systems and war planes. Conservationists have objected on the grounds that the mining will upset the ecological balance of the area and that restoration will be impossible. However, until recently large areas around St Lucia were already closed to the public because they constituted a missile-testing site. A large area around Arniston is closed to the public for similar reasons. The SADF is the largest landowner in South Africa outside of the gameparks and much of its land is devoted to testing weapons and training people for death and destruction. Conservationists must raise their voices of protest about this as well.

The SADF is the largest landowner in South Africa outside of the gameparks

Rainbow coalitions and community empowerment

Green politics calls for an end to the arms race and the process of militarization; it demands the total transformation of our society. Trade unions have a crucial role to play in this process. Workers are often at the frontline of environmental abuse, and so there is much concern to get environmental issues onto the agenda of the labour movement. The intervention of the National Union of Metal Workers of South Africa (NUMSA) in the St Lucia mining controversy is an important development. So is the decision of local trade unions such as the Chemical Workers' Industrial Union (CWIU) and the South African Chemical Workers' Union (SACWU) to take up issues relating to toxic waste. (See Chapter 6.)

Recent protest against the importation of toxic waste by Thor Chemicals brought together a heterogeneous group of people including black peasant farmers, ecologists, students, unionized workers and a traditional Zulu chief. CWIU General Secretary, Rod Crompton, said that the Natal demonstration signalled the emergence of a powerful 'rainbow coalition' between green groups and the trade union movement in South Africa. 'There is a growing realization among the leadership of our union that health and safety cannot stop at the factory gate. What Thor shows so clearly is that what happens inside the factory impacts severely on the community that lives around it' (*The Weekly Mail*, 20 April 1990).

This rainbow coalition is also evident in the recent cooperation between the Dolphin Action and Protection Group (DAPG) and the Food and Allied Workers' Union (FAWU). DAPG's focus on gill netting by Taiwanese tuna fishermen working in South African waters revealed the appalling conditions under which their South African crews were working. (See Chapter 13.)

These alliances between environmental groups and trade unions have been welcomed by representatives of both COSATU and NACTU (Environmental Film Workshop Conference, Johannesburg, 3 November 1990). Unionists have emphasized the importance of 'the right to know' and 'the right to decide' as central to trade union's struggle against the destruction of the environment. As the International Federation of Chemical, Energy and General Workers' Union (ICEF) has stated, 'Information and control are the twin pillars upon which rests the democratic operation of society. Trade union demands for expanded industrial democracy stand on the same principles. Such demands imply and include the assertion of common ownership over the natural environment, rights which have been violated in the past by industry without discussion or consultation' (quoted by Coleman, 1990: 68).

A process of empowerment is also evident in communities which are beginning to challenge the authoritarian tradition whereby game reserves were established without any consultation with the local people. For example, in the Richtersveld, local people have recently secured the legal right to continue to use a proclaimed national park for grazing and to participate in its management. This community mobilized because they said, 'We want to leave a legacy for our children, not crumbs'. Their words are a concrete expression of the principle of sustainable development.

Their example — and that of the Purros project in the Kaokoveld

WELMA ODENDAAL

ABOVE

On 1 May 1990 about 7 000 people marched in support of the residents of Matjieskloof in Namaqualand, who launched a boycott in protest against the cost of services after the village was taken over by the Springbok municipality

in Namibia — point to the way forward, a way where the development of people and the protection of plants and animals can coexist; a way where conservation is not the exclusive domain of the privileged at the expense of the local population.

Another important development is the establishment of progressive environmental groups such as Earthlife Africa. While the latter does not define itself as a political organization the 1990 National Congress adopted a 'statement of belief' which included a commitment to 'grassroots democracy, to ending exploitation and abusive power relations and a rejection of all forms of discrimination'. Earthlife Africa's social philosophy is similar to that of the green movement (Capra and Spretnak, 1984; Papadakis, 1984) and its commitment to non-violent direct action echoes that of Greenpeace.

However, it is not only the young, white, middle-class and well-educated who appear to make up the social composition of Earthlife Africa — or organized workers or directly threatened communities — who are taking up environmental issues. There are grassroots struggles which include environmental issues. A grassroots environmental movement existed in embryonic form in the period between 1984 to 1986, the days of 'people's power'. Through street committees a great deal was done to organize garbage collections and establish people's parks with small rockeries and colourful painted tyres in many open spaces in townships throughout the country (*The Weekly Mail*, 9 May 1986).

'We want our children to have more than just crumbs'

— People of the Richtersveld

Apartheid and the environment

At the grassroots level these efforts have always been a small part of a much broader struggle against apartheid. The President of the Soweto-based National Environmental Awareness Campaign (NEAC) has emphasized that apartheid is at the core of environmental degradation. 'Blacks have always had to live in an environment that was neither beautiful nor clean. We have not had the proper housing, roads or services because the authorities would not accept that we were a permanent part of the city scene.' In 1987 NEAC members demonstrated with placards saying, 'Protect our environment by removing the Group Areas Act and the Land Act' and 'Apartheid makes our townships dirty' (*The Star*, 6 June 1988). It is also apartheid which for years has excluded the majority of South Africans from some of our most beautiful beaches, nature trails and mountain peaks.

In April 1990 an ANC leader, Andrew Mlangeni, was guest of honour at a tree planting ceremony in Soweto arranged by NEAC, 'to highlight the fact that there is a growing environmental awareness among resistance groups' (*The Weekly Mail*, 15 April 1990).

This awareness is also illustrated by the resolution passed by the South African Youth Congress (SAYCO) in April 1990. The resolution blamed apartheid for the 'deliberate destruction of our land and environment' and vowed to fight for a greener South Africa.

The ANC is at present in the process of formulating its environmental policy. According to the head of the Internal Leadership Core of the ANC, Walter Sisulu, 'The environment today is a top priority all over the world. Particularly here. I would say that our country is among those countries that need to pay attention more than others' (interview reported in *New Ground*, vol 1, no 1, September 1990).

In 1989 the Head of the ANC's Department of Economics and Planning, Max Sisulu, in response to a set of questions, indicated the ANC's thinking on environmental issues.

He attributed environmental degradation to apartheid. 'The

ANC believes that a rational ecological protection policy requires the dismantling of apartheid. Widespread overgrazing, soil erosion and serious land degradation in the so-called 'homelands' constitute the inevitable destructive consequences of apartheid. These cannot be reformed or rehabilitated by land-use management measures without first dismantling apartheid. The ANC believes environmental reconstruction constitutes a major task of a free and democratic post-apartheid South Africa. Indeed, an environmentally conscious society can only exist in a free democratic political environment.'

The statement said the ANC was deeply concerned about high levels of atmospheric pollution in South Africa's industrial areas as well as signs that some of the country's major rivers have been seriously affected by industrial toxins. 'In developing its policy on environmental pollution the ANC will be guided by the principle of seeking appropriate means of introducing environmental management into industrial development so that technical innovations also address sources of pollution, its prevention and control ... Industry should look at alternative fuel-energy sources like solar energy to reduce the need to burn coal and other fossil fuels.' The statement

BELOW
A 'people's park' near Braklaagte, created by people removed from Magopa to Onderstepoort

PAUL GRENDON

went on to say that the ANC was strongly opposed to the importation of toxic waste and to nuclear power. 'The usual environmentally-couched argument that nuclear power generates electricity without emission of air-polluting carbon dioxide and sulphur oxides cannot stand the test of vulnerability to nuclear hazards and risks.'

Sisulu also emphasized the ANC's opposition to nuclear weapons. 'South Africa's nuclear research is a flagrant violation of the declared policy of African nations, endorsed by the United Nations General Assembly, to respect Africa as a nuclear-free zone.'

Sisulu said the ANC is committed to protecting wildlife resources and strongly opposed to the illegal poaching of rhino horn and elephant tusks. 'The ANC supports the policy approach of some of the free southern African governments which advocate full community participation in the management of wildlife resources and in the economic benefits flowing from this resource ... if wildlife is perceived as an asset by the communities where it exists, they will take it upon themselves to protect it and not be party to poaching and smuggling.' Finally, Sisulu said the ANC was strongly committed to combining economic development with environmental protection.

These points were developed by Stan Sangweni, speaking on behalf of the ANC at the Environmental Film Workshop conference in Johannesburg in November, 1990. He emphasized that 'the ANC is committed to conservation and the rational use of our natural resources for the benefit of the present and future generations. ANC's position is that we, the present generation, have a responsibility which we owe to future generations of South Africa to preserve the environment for them so that they will find it in a viable and usable form. It is also ANC's position that in planning and implementing an economic growth programme a correct strategy is to maintain a healthy balance between economic and social benefits on the one hand and environmental protection on the other hand. In policy terms this means sustainable development and growth with environmental protection.'

At the same conference Barney Desai, on behalf of the PAC, underlined that 'the environmental issue, is, in fact, an issue of survival and should be an integral part of the organization's overall policy.' He proposed that the PAC's environmental policy should be to adopt a holistic approach; commit itself to the conservation of ecosystems and biotic diversity; and accept the concept of sustainable development, 'with the understanding that a prerequisite for sustainability is that disparities in social, economic, and political development be ended.'

Many South Africans view environmental issues as white middle-class concerns

Most speakers at this conference emphasized the importance of building a mass environmental awareness. At the Conference for a Democratic Future in 1989 Earthlife Africa called 'on the Liberation Movement to campaign wherever possible for the protection and rehabilitation of the environment.' Although this motion was passed, giving environmental issues a central place in the struggle to transform South Africa will not be an easy task. With inflation and unemployment rising, workers are often too concerned with jobs and wages to relate to issues of health and safety in working conditions, let alone the wider issues of environmental health. (See Chapter 6.) At present 60 per cent of our people live in poverty and we have one of the most unequal distributions of income in the world. There will be enormous pressures on a new South African government to create wealth to provide our people with clean water, housing, jobs, health and education. It may be tempted to do so at the expense of the environment.

> **'The greening of our country is basic to its healing'**
>
> — Albie Sachs

Green rights

However, there is growing debate around the idea that environmental considerations should be incorporated in the new South African constitution. ANC constitutional committee member Albie Sachs has received much support for his idea that the constitution should include 'third generation' or 'green rights'. Ironically, these rights were contained in the draft of the 1989 Environmental Conservation Act in the provision that 'every inhabitant of the Republic of South Africa is entitled to live, work and relax in a safe, productive, healthy and aesthetically and culturally acceptable environment.' As Glazewski writes, 'It is unfortunate that this bold approach was not adopted in the final document' (Glazewski, 1989:873).

In November 1990 the ANC Constitutional Committee issued a working document entitled 'A Bill of Rights for a new South Africa'. Article 12 deals with 'environmental rights.'

1. The environment, including the land, the waters and the sky are the common heritage of the people of South Africa and of all humanity.
2. All men and women shall have the right to a healthy and ecologically balanced environment and the duty to defend it.
3. In order to secure this right, the State, acting through appropriate agencies and organs shall conserve, protect and improve the environment, and in particular:
i. prevent and control pollution of the air and waters and degradation and erosion of the soil;
ii. have regard in local, regional and national planning to the maintenance or creation of balanced ecological and biological areas and to the prevention or minimizing of harmful effects on the environment;
iii. promote the rational use of natural resources, safeguarding their capacity for renewal and ecological stability;
iv. ensure that long-term damage is not done to the environment by industrial or other forms of waste;
v. maintain, create and develop natural reserves, parks and recreational areas and classify and protect other sites and landscapes so as to ensure the preservation and protection of areas of outstanding cultural, historic and natural interest.
4. Legislation shall provide for cooperation between the State, non-governmental organizations, local communities and individ-

NAMAQUANUUS

ABOVE
The Hanekoms are among several goat-farming families in Narap who have been given notice by Gold Fields to make way for a gamepark. Community farmers between Springbok, Okiep, Concordia, Hoits and Carolusberg are being threatened with eviction by the mining company, who own extensive tracts of land in Namaqualand

uals in seeking to improve the environment and encourage ecologically sensible habits in daily life.

5. The law shall provide for appropriate penalties and reparation in the case of any direct and serious damage caused to the environment, and permit the interdiction by any interested person or by any agency established for the purpose of protecting the environment, of any public or private activity or undertaking which manifestly and unreasonably causes or threatens to cause irreparable damage to the environment.

In the final analysis the best insurance for the future is a high level of mass environmental awareness. It is clear that such environmental awareness is growing in South Africa. But this must be part of a wider emphasis on community participation and the democratization of social and political life. Grassroots democracy is central to green politics. The philosophy behind green politics is that of 'deep

ecology'. This denies that human beings are separate and superior from the rest of nature. This ecological consciousness is in sharp contrast with the dominant world view of technocratic-industrial societies which have become increasingly obsessed with the idea of dominance: with dominance of humans over non-human nature, masculine over feminine, the wealthy and powerful over the poor. Deep ecology is against dominance and for equality. It is not anthropocentric — it does not only focus on our own species. Earlier this century Albert Schweitzer noted,

> It was once considered stupid to think that coloured men were really human and must be treated humanely. This stupidity has become a truth. Today it is thought an exaggeration to state that a reasonable ethic demands constant consideration for all living things (Tobias, 1988:177).

'We must live simply that others may simply live'

— Schumacher

In South Africa we have an urgent struggle to overturn such 'stupidities' and 'exaggerations'. 'Think globally, act locally,' is a key theme in this struggle. We are establishing global solidarity in the struggle against environmental abuse. At the same time there are many local grassroots campaigns developing. They are focusing on a wide range of environmental issues ranging from asbestos dumps to water and air pollution; from the ivory trade to soil erosion; from the promotion of lead-free petrol to the problems with herbicides in Natal; from saving Chapman's Peak to local refuse collection; from nuclear energy to toxic waste. This concern with protecting the environment provides an effective way for all South Africans to invest their energies and unite towards achieving a greener South Africa. Such efforts have both a unifying and healing potential. As Albie Sachs said recently,

> Just as apartheid penetrates through to every aspect of South African life, so must the struggle against apartheid be all pervasive; it is first and foremost a battle for political rights, but it is also about the quality of life in a new South Africa. It is not just playing with metaphors to say that we are fighting to free the land, the sky, the waters as well as the people. Apartheid not only degrades the inhabitants of our country, it degrades the earth, the air and the streams. When we say Mayibuye Afrika, come back Africa, we are calling for the return of legal title, but also for restoration of the land, the forest and the atmosphere: the greening of our country is basic to its healing ... There is a lot of healing to be done in South Africa (Sachs, 1990).

•CHRIS ALBERTYN•

'I'm a flexible green,' says Chris Albertyn, a touch obscurely. What he means, he explains, is that his personal philosophy is 'dark green' but that the radical language of dark green is inclined to frighten off newcomers to the environmental movement — and that would be counter-productive.

Albertyn is a youngish sprig in what is, in world terms, a relatively late-blooming green movement, but this psychologist from Pietermaritzburg has certainly made his presence felt — to the considerable annoyance of many Natal industrialists.

Trips into the country with his woolbroker father and a 'really great' biology teacher inspired in Albertyn his concern for nature, an interest intensified by the choice of zoology as a co-major at university. But it was his reading in his chosen professional field — psychology — that set Chris Albertyn on a new train of thought. 'Fritjof Capra's *The Turning Point* had a profound effect on me. It looked at the implications of new physics to occupations other than physics. It looked at relationships between things rather than at things in isolation.' His training as a family therapist also broadened Albertyn's viewpoint, forcing him to 'think about things ecologically and systemically'.

Environmental issues always hovered on the edges of his consciousness and, during a stint as a sub-editor on the *Natal Witness* after he had finished his internship in 1987, he began to realize, through the influx of wire service material, how important environmental issues were becoming abroad. Locally, though, they were being pushed aside by editors who believed the public wasn't interested.

Albertyn was interested, though, and found himself thinking and reading more and more about these issues. 'I was feeling quite alienated. There I was treating sixteen-year-old children who were wetting their beds and the world was collapsing and nobody was taking any notice.'

Albertyn feels that society at large is more inclined to put its energies into dealing with family and personal problems than into attention to broader systemic problems. 'Both social and environmental destruction are part of the problem. It's where you intervene in a system that counts.'

Where he chose to intervene was in the area of industrial pollution. 'There are all these multinationals with slick PRs and the public sit back and accept the line they are fed. There wasn't a powerful voice accountable only to the environment.'

At the start of 1989, Albertyn, who had been peripherally involved in the work of the Society Against Nuclear Energy (SANE), was introduced to the work of Earthlife Africa ('which was basically a reading group in Johannesburg') and agreed to launch a branch in Pietermaritzburg. Though he was registered as a psychology masters student, Earthlife quickly became a fulltime job.

'Issues just snowballed. The more I read the more incensed I became. I decided we had to inform the public.' The campaign which shot Earthlife into the public consciousness via the media was directed against the quantities of toxic waste polluting Natal's rivers and poisoning the communities dependant on those rivers. 'It's always the politically or economically disempowered who suffer at the end of the day,' observes Albertyn.

In planning his campaigns, Albertyn made a point of drawing in as many people as possible. He saw Earthlife as an invaluable opportunity to involve in social issues people who had not been involved in other social responsibility areas and 'through that to promote a broader consciousness of the interconnectedness of things.'

It evidently worked. Within six months there were more than 700 members in Natal and a branch was launched in Durban. Fifteen months after the Pietermaritzburg launch there were fifteen branches of Earthlife in Southern Africa. Albertyn found himself increasingly in the limelight — something which he

could happily have done without. 'The more confident I got, the more I learnt and the higher the profile became, which was a problem.'

With each campaign came new adherents. The issue of hormonal herbicides brought in vegetable farmers. The protest over Thor Chemicals, where workers were suffering the effects of mercury poisoning and effluent was polluting the rivers, brought together black trade unionists, rural peasants and white farmers.

'In some senses what we saw was that we have drawn in people in Natal who would never have become involved in human rights activism at all; people who have become a lot more tolerant, who have questioned and, despite the fact that they accept that these issues can be seen as highly political, still remain a part of the organization.'

Earthlife, Albertyn hastens to add, does not see itself as a political organization. It does not even see itself as an organization with a fixed agenda.

'I still think it's a movement because we haven't formulated clearly our identity. I would like to see it becoming a facilitative, grassroots organization which people, particularly communities who want to take up an issue, can get hold of to tap the expertise of our membership on a voluntary basis.

'At another level, I would like to see it organizing people to take up issues and create pressure on behalf of communities which don't have the voice and the resources.'

On a personal level he tries to follow a 'green lifestyle'. But he admits that it is not easy and there are many contradictions.

What he has not become is dogmatic. He works hard to avoid becoming 'prohibitive' and 'more holistic than thou', particularly when writing. 'We've got thoroughly involved in putting out a newspaper and spreading information and we have found it very difficult to be balanced and avoid the moral high ground.' Ever the psychologist, he is only too aware that 'If people start doing things out of guilt it doesn't work.'

While individual conversion to a greener lifestyle is important, Albertyn believes the movement needs to extend its work far beyond that. 'We have to get our views over to the politicans, the decision makers. There's a definite need for activism and that is what I have been trying to promote.'

With people being as important to him as nature, and with the example of some of the European green movements to warn him, Albertyn is well aware of the dangers of imposing solutions without consulting those whom they will most seriously affect. He is aware, too, of the urgency of the situation and struggles with the problems of trying to maintain a healthy balance between the need to consult and the need to act.

'The right thing to do in terms of the environment could be socially destructive, and that's not going to solve any problems at all. We must be careful of eco-fascism. We must get a participatory decision.'

There are no clear-cut answers and Albertyn confesses to being 'confused'; to suffering bouts of cynicism and disillusionment. But 'then the anger takes over and you start up again'. ❐

2

RAINBOW ALLIANCES

Community struggles around ecological problems

◆ EDDIE KOCH ◆

In October 1990 a man from Makurung, a little and remote village in the far northern Transvaal, phoned the offices of *The Weekly Mail* newspaper with an unusual request. His neighbours, said the man, lived next to a dump of waste left by the local platinum mine. During the day, when the wind blew, the dust got into people's noses, irritated their eyes and created an abrasive layer between their clothes and skin. Overnight it filtered into their homes and left behind a fine patina on the floors, the furniture, the stove, and anything else that had been left uncovered.

The people of Makurung had lived with the dust for years, regard-

BELOW
Children in Ntambanana in the north-eastern Transvaal

GAVIN YOUNGE

ing it as an irritant and a nuisance but accepting it as part of their natural environment. Recently, however, a nearby village called Mafefe, located amidst the rugged Steelpoort Mountains some 80 kilometres to the north-east of Makurung, has received extensive publicity about abandoned dumps of blue asbestos dust that were playing havoc with the lungs and lives of the people who lived there. (See Chapter 3.)

'Now we are not so sure about this mine dump,' said the caller. 'We need to know if there are any poisons in its dust and how it is affecting us and our children. We think you should come up here and investigate.'

Community involvement in environmental issues

The telephone call from the far northern Transvaal signalled the way in which people across South Africa, ranging from nomadic pastoralists in the distant reaches of Namaqualand to ratepayers in plush seaside resorts, had by the end of the 1980s become aware of the problems facing the environment around them — and had begun to organize against the businessmen and bureaucrats whose actions were degrading the quality of the environment and thus of their everyday lives.

At the end of 1989, my newspaper's regular review of the state of the country's ecology could find only one example of community-based organization around the environment: the campaign at Mafefe. The Mafefe committee has become legendary for the way it has rallied local people around a campaign to monitor levels of asbestos dust in the air, survey the extent of asbestos-induced lung diseases in the village, and agitate for effective remedies for the eco-disaster inflicted upon the community.

By the end of 1990, things had changed.

Heaven and hell: Zamdela

Take the township of Zamdela. It was built downwind from Sasolburg, perhaps the most industrialized and polluted city in South Africa, so that white suburbs in the area could be spared the full blast of soot, carbon dioxide, nitric gases and other noxious matter that spews from the stacks of Sasol's outdated oil-from-coal plant. This plant is said to be among the most environmentally-damaging factories in the world, while the nearby Lethaba power station is the largest of Africa's coal-burning power stations. There are also a number of steelworks and chemical plants nearby.

'When I first arrived in Sasolburg it was three in the morning. It was beautiful, all those lights. I thought I was in heaven,' says factory worker Tladinyane Kgodumo. 'Now, after all these years, it seems more like hell. Workers in these factories are fighting a constant struggle for better health and safety regulations. They work all the time with dangerous chemicals and there are many accidents. But it is not only the factory workers who suffer from working with these chemicals. What about the people who live in Zamdela, right next door to the factories? We are all suffering from the pollution

We need to solve our socio-economic and political problems before people will be receptive to environmental issues

these factories have been putting into the air.'

Today Kgodumo and his compatriots in the South African Chemical Workers' Union (SACWU), an affiliate of the National Council of Trade Unions (NACTU), are active members of a small and struggling environmental pressure group that has been formed in the township.

The local township doctor says children suffer from allergy-related problems such as asthma, itchy palates, streaming noses and infected eyes. Many cars in the area have corroded roofs and the stench that pervades the air is so severe that, says resident Abraham Mahlong, 'if you drive through Sasolburg in a taxi with strangers, they all look at each other.'

The Zamdela environment group is busy planning a survey of air pollution, its extent and its effects on people's health, as the first stage of a campaign to pressurize local industry into improving their living conditions. But workers do not have access to laboratories and sophisticated monitoring equipment, so they have been advised to use another device: to place buckets of water outside their homes at night and, when they wake in the morning, to measure the thickness of the layer of soot that has settled on the surface.

The township's green activists, backed by their union, have engaged the assistance of scientists from the University of the Witwatersrand, who will help analyse the samples.

During 1990, each month seemed to spawn a new case in which ordinary men and women strove to correct the wrong that had been done to the water, the air, the land or the rivers around them. Often the publicity given to each of these struggles would inspire others to do battle against the forces spoiling their environment.

Problems in paradise: Maputaland

In the Maputaland district of northern Natal, an environmentally unique region that runs along the southern border of Mozambique and South Africa, another set of ecological struggles has been dominating people's lives. The KwaZulu government, which administers the area, is planning to consolidate a string of nature reserves and game parks in Maputaland and, aided by R50 million granted by the Gencor mining corporation, hopes to create a conservation area that will rival the Kruger National Park in size.

The problem is that thousands of Tembe-Thonga people have inhabited the area for centuries, developing a lifestyle that relies on fishing, harvesting the wild fruit trees that abound on the semitropical coastal plain, growing bananas in the swamp and marshlands near the coast, and gathering roots and herbs for medicinal purposes. State-sponsored conservation has led to severe social and geographical dislocation; forced removals; material loss through inadequate compensation for land; restricted access to lakes for fishing and fields for farming; electrified game fences that separate villages from their fields; and controls on the collection of thatching grass, trees, fruits, roots and herbs.

These policies have provoked a mood of sedition in Maputaland.

In August 1989, a dozen informal leaders from the area met the Congress of Traditional Leaders of South Africa (CONTRALESA), an alliance of traditional chiefs who align themselves with the ANC, to discuss their problems. Three months later a delegation of tribal elders organized a movement called 'Isididi' to mobilize against the Kosi Bay nature reserve and the restrictions that local conservation measures were imposing on communal lifestyles.

As part of this groundswell of resistance to KwaZulu's conservation policies, a group of school children from Maputaland workshopped a play which they staged at schools in the Durban area to highlight their plight. The performances were cancelled when Nick Steele, director of KwaZulu's Bureau for Natural Resources, lodged strenuous objections to the drama.

Conservation and the people: KaNgwane

The KaNgwane government, led by former Chief Minister Enos Mabuza, adopted a different approach to conservation in the eastern Transvaal homeland. Traditional conservation, which concentrated on the creation of protected areas and the preservation of single species in isolation from Africa's social problems, would not survive: so said Mabuza at the launch of a new programme to preserve South Africa's natural heritage in December 1990. Humankind would not revere the air they breathe and the land they inhabit if bound by poverty and dehumanizing conditions. 'We need to solve our socioeconomic and political problems before people will be receptive to environmental issues.'

The breathtakingly beautiful Mthethomushwa and Songimvelo nature reserves were established in KaNgwane during the course of 1990 through consultation with the local chiefs and their tribal councils, and with their consent. Profits from tourism are shared with the communities living alongside the reserves and local residents are allowed into the protected areas to collect grasses, herbs and roots.

'I suppose we are talking about rural development,' says head of KaNgwane Parks' communications section, Karl Lane. 'People want to escape the trap of rural poverty and they see the way out through jobs and education. So we must create jobs and generate profits which can be used to the benefit of the community, such as building schools.' The KaNgwane government is currently sharing with the local people the profits made from tourism to the reserves, and sometimes the communities receive up to 60 per cent of the revenue generated.

Acknowledging the community: Purros

Margaret Jacobsohn, who helped set up a similar and highly successful conservation-cum-development programme at Purros in the Namibia's Kaokoveld region, describes in her contribution to this book how Himba men and women, who previously viewed conservationists with hostility, have been mobilized to play an active part in running the new reserve. (See Chapter 14.)

She points out that the programme, widely acknowledged as a model to be emulated throughout Africa, had to alter social relations

The KaNgwane government is currently sharing with the local people the profits made from tourism to the reserves

and create new institutions in the community in order to ensure full participation in the scheme and equitable distribution of the revenue generated by tourism. Money is paid directly to the village elders and discussions about how the amounts are to be distributed are held at communal meetings. Women are encouraged to attend because they have 'a large input into family decisions and as the distributors and de facto controllers of food and other goods they wield considerable domestic power'.

Flowers in the desert: Namaqualand

David Fig's chapter in this book (see Chapter 8) provides a vivid description of the struggles mounted by pastoral communities who have inhabited Namaqualand, a remote and overlooked wedge of territory between the western Cape and Namibia, for centuries. Their struggles have been against conservation programmes that have deprived people of their land and grazing rights, against plans by ESKOM to build a nuclear power station in the region of Komaggas, and against the creation of one of the world's only storage sites for nuclear waste in the midst of the villages where the people live.

These multifaceted struggles by people of whose existence many South Africans are not even aware generated some unusual forms of protest.

About 6 000 people marched under the black, green and gold banner of the ANC across a landscape of semidesert to protest at the proposed construction of a recycling plant at Alexander Bay for imported toxic waste. A message of solidarity from the Pitjantjatjara people of the Uluru-Katajuta National Park at Ayers Rock in Australia was sent across the seas to the people of the northern Richtersveld, expressing support for Namaqualanders' struggle against being expropriated in the name of conservation. The newly-formed Namakwalandse Burgersvereniging spent much of its time consulting botanists, social anthropologists, nuclear experts, energy researchers, environmental lawyers and land-use experts to devise programmes to defend their environment.

'Why all of a sudden is there all this activity and protest to save animals when there was no reaction at the time when people faced removal?'

— Mike Mabuyakhulu

Toxic waste: Azaadville

Earthlife Africa, which made a spectacular entry as South Africa's newest environmental group in 1990, began working with left-wing civic organizations as well as the more established (and discredited, in the eyes of activists) community council of Azaadville, a segregated township for Indians near Krugersdorp. The project was to oppose a plan by local businessman Benoni van Graan to build a disposal site nearby for the toxic waste generated by local industry.

The people of Azaadville appeared to be so aghast at having to live next to a poison dump that they mended the township's internal political tensions and even made tentative approaches to the ultra-right-wing Conservative Party, whose members in the white suburbs of Krugersdorp were opposed to the idea, to discuss joint protests.

ROB MEINTJES, NAMAQUANUUS

ABOVE

Farmers in the northern Richtersveld at a meeting in Khubus in September 1989 to challenge the state, which wanted to declare the Richtersveld a national park without the consent of the communities

An explosive situation: Rooi Els

White ratepayers, not renowned for their militancy, also mobilized themselves in Rooi Els, a little seaside resort near Cape Town, to protest at the activities of a subsidiary of Armscor.

The company, known as Somchem, was testing rocket propellants on a nearby test range that sent large clouds of dust and chemicals into the air above the local dam. This created a fear among residents that their water supplies were being contaminated.

Black, white and green: Thor Chemicals

Late in 1989 journalists discovered that large quantities of mercury, red-listed as one of the most dangerous known toxins, were leaking from the plant of the Thor Chemicals factory in the Natal Midlands. Analysis of water and sediment from the Umgcweni River showed some of the highest mercury contamination levels ever recorded, placing a nearby squatter community at serious risk.

In April 1990 protests were launched by the Earthlife branch in Pietermaritzburg in cooperation with CWIU and Greenpeace International. The protests were directed not only against Thor, but also against the US multinational, American Cyanamid. The American corporation exports 10 tons of waste to the South African plant each year. Black peasants, white farmers, students, unionized workers, green activists and a traditional Zulu chief banded together to protest at the importation of toxic waste to South Africa and held a large demonstration outside the gates of Thor.

Rod Crompton, general secretary of the CWIU, commented that the protest signalled the potential for a powerful 'rainbow alliance' between green groups and the country's black labour movement. 'What Thor shows so clearly is that what happens inside the factory impacts severely on the community that lives around it.'

Herbicides: Natal farmers

The white farmers who joined the protests at Thor had been conscientized by ecological problems of their own. In 1987 rain samples taken in the vegetable fields of the Tala Valley near Pietermaritzburg revealed concentrations of the herbicide 24-D that were one million times higher than the amounts capable of damaging vegetable crops. The rain also contained levels of a similar chemical, 245-T, that were 10 000 times higher than the legal limit in the United States. Both chemicals became notorious after they were used in the manufacture of 'Agent Orange', the defoliant used by the US Army to destroy vast tracts of the forest that gave cover to communist guerrillas during the Vietnam war.

The Natal Fresh Produce Growers' Association, whose members claimed to have suffered millions of rands' worth of damage to their crops as the herbicides drifted onto their farms, launched a massive lawsuit against twenty local and multinational chemical companies. The application in the Pietermaritzburg Supreme Court sought an interdict against the sale and manufacture of all such herbicides in the country. The case was rejected on a technicality early in 1990 and the consortium of chemical companies threatened to claim costs in the region of R1,5 million unless the farmers retracted and agreed to sign a humiliating acknowledgment that there was no proof that the herbicides cause environmental damage. The gutsy farmers refused to buckle under the pressure and vowed to prepare another application, even though some of them may have to sell their farms to pay the legal costs.

Double destruction: Taiwanese trawlers

Another show of united action between diverse groups and organizations was staged in mid-1990 when the Food and Allied Workers' Union (FAWU), a militant union with a leadership loyal to the South African Communist Party, joined forces with Earthlife's Cape Town branch and the Dolphin Action and Protection Group to protest at the activities of Taiwanese fishing trawlers. The owners of these ships were accused of stripmining South African waters by fishing with illegal gillnets. (See Chapter 13.)

Nature has a magnetic power to rally divergent, and sometimes even antagonistic, groups in its defence

FAWU was prompted to join the protests by revelations that at least four South African workers employed on the trawlers had had their fingers amputated because of gangrene and frostbite caused by working without adequate protective clothing in sub-zero temperatures in the refrigerated holds of the ships for shifts of up to eleven hours. 'The Taiwanese use cheap South African labour then come into harbour, dump the workers and scoot,' said Cape Town attorney Sandy Liebenberg.

Dagga and defoliants: paraquat

Just before Christmas in 1990, a 'community' of a different kind won a small green victory. South Africa's many thousand smokers of cannabis, known locally as 'dagga' and used by some as a traditional herb, were able to breathe more easily when Police Minister Adriaan Vlok agreed to suspend a programme aimed at wiping out the nation's dagga crop by spraying fields from the air with a deadly defoliant called paraquat.

Vlok made this decision after holding talks with Earthlife Africa's southern Natal branch and the South African Rivers Association, who claimed the government was waging 'chemical warfare' on rural villagers who live near the dagga plantations.

Community groups are frequently denied access to information that is vital for environmental campaigns

Paraquat is registered here as a class one poison and is responsible worldwide for more poisonings than any other weedkiller. Tiny amounts of paraquat can be lethal if swallowed, absorbed through the skin or inhaled. Toxicologist Robert Dreisbach says in his *Handbook of Poisoning* that amounts as small as one microgram (one hundredth of a milligram) can cause fatal poisoning.

It is restricted to licensed users in the United States and is banned in or severely restricted in Sweden, Denmark, the Phillippines, New Zealand, Turkey and Israel. The weedkiller is also listed as one of the 'dirty dozen', the twelve hazardous pesticides and herbicides that green groups are campaigning to have abolished worldwide.

'Paraquat is an environmental killer and public hazard of the worst kind. We are concerned for the safety of the rural population and the effect of paraquat on the environment, wildlife and rivers,' said a joint statement from the Natal branch of Earthlife Africa and the SA Rivers Association.

The two groups said they were particularly concerned for the health of villagers who live in the areas being sprayed and for small-scale farmers whose crops were likely to be damaged in the campaign. Dagga crops are frequently planted among or close to other products, especially maize, so that they can be concealed from the police. The two organizations agreed in their talks with Vlok to investigate environment-friendly methods of dealing with dagga as well as ways in which to encourage people who rely on cannabis cultivation for a living to explore other kinds of farming, including the creation of herb-growing schemes.

'We put forward the idea that the dagga trade needed to be understood in a broader context,' said a joint statement by the groups after their meeting with Vlok. 'It can never be solved until the growers are given a viable cash crop alternative and access to markets. Consultation with the people most affected by these programmes is essential if any intervention is to succeed.'

A press statement issued after the talks with Vlok noted that one of the most powerful levers used to persuade him was a threat to bring legal action for damages on behalf of small black farmers who suffered crop damage because of the herbicide.

'Mayibuye Afrika': The land issue

The above are only the most explicit examples of action by diverse groups and local communities to defend their environment. Other protest and resistance campaigns have contained within them the seeds of a green awareness. The land issue, one of the most pressing problems for the government and opposition groups during this period of transition in South Africa, is intimately bound up with ecological issues.

David Cooper, in Chapter 12 of this book, points out that black peasants frequently do not have an immediate sense of grievance about the dongas and sheet erosion that have scoured the land they have been forced to live on.

However, once they begin to discuss the issue rural people often become crucially aware that this problem, perhaps South Africa's greatest ecological crisis, is linked to the legacy of dispossession, forced removals, discriminatory allocation of resources and rural segregation that characterized apartheid.

'When we say Mayibuye Afrika, come back Africa, we are calling for the return of legal title, but also for the restoration of the land, the forest and the atmosphere; the greening of our country is basic to its healing,' says ANC legal expert Albie Sachs.

'The establishment of people's parks during the uprisings of the early 1980s was more than the creation of leisure areas, it was an affirmation of the desire for beauty in the community. Prisoners on Robben Island planted flowers and grass wherever they could, even if the result was to prettify their dungeons in the eyes of visitors.'

A rainbow of alliances?

The assortment of community struggles that have begun to blossom across South Africa demonstrate one overriding feature: nature's magnetic power to rally divergent, and sometimes even antagonistic, groups in its defence. Said Earthlife representative Chris Albertyn after the Thor demonstrations: 'It seems that people who are opposed to each other on other issues are prepared to work together to save the environment. This is a force for conciliation, unity and peace in South Africa.'

It seems that people who are opposed to each other on other issues are prepared to work together to save the environment

But rainbows are ephemeral things. They emerge suddenly, casting a brilliant beauty over the landscape, and then fade in an instant. Nobody knows when they will return. The alliances that have emerged around the environment need to guard against the same fate.

The Thor protests were a powerful example of green mobilization. For the first time, a South African company was forced temporarily to close its plant because of the damage it was causing to the local surroundings. But the factory has since reopened, shipments of mercury continue to come into the country and, at the time of publication, this 'rainbow alliance' of unions, farmers and environmentalists had yet to reappear.

The reasons are manifold. South Africa's environmental movement is still in its infancy. The dedicated men and women who run it are overworked, overextended and seldom paid for the work they do. Trade unions have a series of burning issues, both in the workplace and in the political arena, that dominate their agendas. No sooner does one ecological crisis resolve itself, but another is upon us.

After Earthlife and Greenpeace succeeded in shutting down Thor, there were reports that even more ominous consignments of poisons were on the verge of being sent to 'homelands' like the Ciskei and the Transkei. These reports required intensive research and investigation, leaving little time and energy to maintain the pressure on Thor.

The need for a permanent environmental organization

Clearly the time has come for the green groups that have proliferated in the last few years to consider the creation of a permanent organization with staff and facilities to ensure continuous and efficient operations. The Richtersveld communities were able to call on the support of expert advisers to help them in the formulation of green tactics. The people of Makurung have not yet received this kind of assistance. A well-staffed and skilled organization that can consult community groups and offer skilfully researched advice on ecological issues is crucial for the success of community-based green struggles.

There is also the need for a monitoring group that is capable of maintaining a consistent and scientific watch over factories, farms and mines that degrade the environment. Such a group should also provide accurate data to activist groups and media so that credible public-awareness campaigns can be mounted. Other countries have statutory bodies that perform this function, such as the Environmental Protection Agency in the United States. In the absence of a similar body here, we need to create our own independent monitoring body.

The need for a free flow of information

There is another compelling reason for an environmental organization capable of conducting rigorous and reliable research. Community groups are frequently denied access to information that is vital for environmental campaigns. There are a host of laws and regulations that prohibit the free flow of this vital data. The Nuclear Energy Act of 1982, for example, contains provisions that resemble the old State of Emergency regulations.

'When the Minister or his delegate is, in the exercise of any power in terms of this Act, of the opinion that it is in the interests of the security of the state that the reasons for his proposed exercise of such power in any direction be not disclosed, he need not disclose those reasons to any person affected thereby,' says one of its clauses.

When *The Weekly Mail* tried to obtain details about toxicity tests and environment impact studies conducted on paraquat before it was licensed for regular use, we were told by the government's registrar of pesticides that the information was confidential and the

AVIGAIL UZI

ABOVE
A Parktown homeowner protests against plans by an oil company to start a factory in the leafy northern suburb of Johannesburg

'Whites see beauty in the flight of birds and grace in the movement of animals. Blacks see a possible source of food'

— Gomolemo Mokae

department was bound by a secrecy clause protecting the commercial interests of the product's distributors.

The people of Namaqualand have had first-hand experience of these clamps on the free flow of information. Elizabeth Beukes works for *Namaquanuus*, a community newspaper that spearheaded the protests against nuclear issues in Namaqualand. After a visit to the waste storage facility of the Atomic Energy Corporation (AEC) at Vaalputs, she wrote: 'We learnt that there was no consultation whatsoever with the people of Leliefontein (an indigenous settlement near Vaalputs) when selecting the site. For the AEC our people did not seem to exist.'

Paul Johnston and Jim Vallette, senior officials from Greenpeace International who attended a conference in Johannesburg on the environment in late 1990, placed heavy stress on the need for environmental democracy. 'People have the right to know the size and destructive potential of poisons discharged into their life support systems,' they said in a report following their visit to South Africa. 'Freedom of information and public participation in government decisions are often ignored tools in the global struggle for environmental rights. We would never have known that the United States ships toxic wastes to South Africa if the US government was not required to give this information under the Freedom of Information Act.'

Given the absence of such rights in South African law, it is vital that environment groups devise ways to investigate and find this data themselves and conduct campaigns that force the authorities to make it available.

The need for unity

The healing potential of environmental campaigns is not always fulfilled, and this is another danger that needs to be guarded against. There are frequently occasions where mobilization around the ecology can create deep divisions. Chapter 6 of this book deals with the danger of a rift developing between unions, whose main task it is to protect jobs, and environmentalists, always keen to shut down plants that pollute the atmosphere or rivers.

Green groups across the world are renowned for attracting activists who possess an enthusiasm and sense of righteousness verging on fanaticism. There have already been examples of splits and factions developing within and between South Africa's green groups. Splits emerged at Earthlife's 1990 national congress over the contentious issues of abortions and animal rights.

The prospect of thousands of seals being culled on the Cape West Coast galvanized a powerful green alliance during 1990 and the unprecedented level of protest and publicity succeeded in having the 'harvesting' programme postponed. But the manner in which it was carried out opened up serious rifts between environment groups. The Wildlife Society of Southern Africa and Earthlife Africa had bitter disagreements over the harvesting of seals. The

PAUL GRENDON

ABOVE
The Namakwalandse Burgervereniging is launched in Steinkopf in July 1989. In addition to its other activities, the organization later took part in a campaign against the dumping of nuclear waste in the region

prospect of a permanent breach between the two was avoided when representatives from each group sat down one night in Johannesburg and after a long discussion realized that in fact they had much in common.

The need for consultation

But while these groups were bickering in public, both were accused by the Azanian Peoples' Organization (AZAPO) of paying more attention to the welfare of animals than to the lives of oppressed people. 'Now at the height of vogue, the green movement agitates on behalf of the flora and fauna of the entire animal kingdom with but one exception: black mankind ... Whites see beauty in the flight of birds and grace in the movement of animals. Blacks see a possible source of food,' said AZAPO Vice-President Gomolemo Mokae.

Earthlife and the Wildlife Society agreed that a key principle for containing internecine divisions was the need to consult local communities over ecological issues that affect them and take guidance from these people in devising strategy and tactics for environmental campaigns. The vital importance of this principle is demonstrated in Jacobsohn's article on the community-based conservation project at Purros (Chapter 14). It has also now become sacrosanct in green circles but is sometimes observed in the breach.

For example, in 1989 South Africa witnessed one of its biggest environmental campaigns when a whole host of organizations agi-

For the principle of consultation to work, it clearly needs more than lip-service to be paid to it. It will also frequently require patient and arduous organization

tated to stop the beach dunes at St Lucia in Natal from being strip-mined for titanium — only, none of them remembered to consult the workers at the company that planned the project. Many of these workers had been forcibly removed in the 1970s from the very dunes that Richards Bay Minerals now wants to mine to make way for the nature reserve at St Lucia.

'We are sensitive to the growing environmental awareness in this country,' said Mike Mabuyakhulu, an organizer for the National Union of Metalworkers of South Africa (NUMSA). 'But there is one question that our members at Richards Bay Minerals are asking and that is: 'Why all of a sudden is there all this activity and protest to save animals when there was no reaction at the time when people faced removal?'

Is it because, this time, there is a threat to the survival of a favourite holiday resort for whites? None of the environmental organizations have consulted us about the issue and some of our members are wondering if these groups think it is more important to save insects and animals while we have to sacrifice jobs and wages.'

For the principle of consultation to work, it clearly needs more than lip-service to be paid to it. It will also frequently require patient and arduous organization.

Jacobsohn says of her work at Purros: 'If conservationists are to put into practice the now very acceptable theory that conservation must involve and benefit local communities, it may be necessary to set up new institutions: new interfaces between nature conservation officials and the people. And here great care and sensitivity must be exercised by the conservation officials ... Any such liaison body must have on it those locals the people themselves choose to represent them.'

The need for constant pressure on the state

The experiences of environmental degradation in the Eastern Bloc, the West, as well as the Third World, have shown that no government, however democratic, can be fully relied upon to protect the environment. Crompton and Erwin point out in Chapter 6 that a far better guarantee is the creation of a powerful network of private groups and organizations within civil society capable of exerting constant pressure on the state to protect their surroundings.

To some extent this network has begun to grow in South Africa and it has already had some phenomenal victories. The Richtersvelders pioneered an agreement that allows them access to the new nature reserve on their grazing lands; asbestos dumps in Mafefe are being rehabilitated; seal harvesting was suspended by the government; Vlok halted the use of paraquat on dagga plantations; stringent controls were imposed on Taiwanese gill-netters.

Fig has likened the environmental struggles of communities in Namaqualand to the carpets of flowers that transform the harsh desert landscape in spring. But these beautiful fields, like some green campaigns, bloom for a few weeks and then die. To permanently transform our environment, community-based programmes need all the sustenance they can get.

3

RISKING THEIR LIVES IN IGNORANCE

The story of an asbestos-polluted community

◆ MARIANNE FELIX ◆

One man's story

'With the disease in my body I have no hope. And my main worry is that I still want to educate my children,' reflects Lucas Maalebogo Maenetja, a forty-four year old man who was told in June 1990 that he has mesothelioma. Mesothelioma is an aggressive cancer caused by exposure to asbestos. Affected people seldom live longer than fourteen months after diagnosis. Since Lucas worked on an asbestos mine as a young man he is eligible for workers' compensation and has received a lump sum of R3 079, an amount equal to four months' earnings. His bitter response to this payment highlights the inequities: 'The money is too little. I worked many years for the mines. Why can't they look after me better?'

Lucas was born in 1946 at The Downs, a farming community in the north-eastern Transvaal, perched on the crest of the majestic Drakensberg. There his father worked on the prosperous white-owned farms. Often his mother sent him laden with vegetables and fruit to visit his aunt in Mafefe, a community of villages nestling in the foothills some ten kilometres away. He looked forward to these errands as Mafefe was an exciting place. Lucas would leave immediately after school on a Friday so that he would have time to linger around the Dalton Asbestos Mill and still be at his aunt's home before nightfall.

Dalton Mill was situated on the outskirts of Ngwaname, the largest village in Mafefe. Lucas was fascinated by the workers disappearing into and reappearing out of the thick clouds of blue-white

BELOW
The deserted asbestos trading store in Mafefe in the northern Transvaal

CEDRIC NUNN

dust emanating from the mill. He would stay for the weekend, playing with his cousins, who had taught him how to swim. Much of his time in Mafefe was spent splashing in the Mohlaphitse River, or making mud oxen on its banks. He remembers how clumps of asbestos fibres clung to the river banks.

These were not the only times Lucas came into contact with asbestos. After he had completed Standard Two his father could no longer afford his tuition, so Lucas found work for two years on Mr Potgieter's farm. It was situated near Bewaarkloof, an asbestos mining community 40 kilometres north-west of Mafefe. Lucas's main task was to deliver milk to the white woman in charge of the white miners' single quarters. Wherever he went in Bewaarkloof he was aware of a layer of asbestos dust.

When Lucas was old enough to work in the asbestos mines he started at Letlakaneng mine in Ramonwane, another village in Mafefe. He must have been about seventeen years old as it was the year 'the rand replaced the pound' (1963). As a novice worker he was responsible for the donkeys that transported the cobbed asbestos down the mountainside. It was difficult convincing the donkeys to move when heavy sacks of asbestos were placed on them. Lucas had to push and shove, even strike them, to persuade them as far as the valley where the mill was situated. There he helped offload the sacks. Long before the end of a day's work both he and the donkeys were covered in a layer of fine asbestos fibres. Wages in the asbestos mine were very low and after two years of strenuous work in the sun he secured a better paying job on the Witwatersrand gold mines.

Lucas was to know no other life than the life on the mines. Since he was a migrant worker, his time at home in The Downs was always short — limited to a few months every year. Nevertheless, he married Maria, who looked after his three children at his home in The Downs. For Lucas migrancy was a frustrating, unsatisfactory way of life, and possibly contributed to his leaving his wife for Sylvia in 1983. Sylvia is now a hawker in Kwa Thema, near the mine where he works, and makes enough for herself to survive on. 'She refuses to touch any of my money as it is little. She says it's better that I maintain my children with that money.' His two sons and daughter, aged twelve to sixteen, are all at high school.

In July 1989 Lucas noticed that he became short of breath on mild exertion and had developed a cough. He reported these symptoms to the mine medical officer who, after taking an X-ray, suggested he have a repeat X-ray when he returned from his leave in October 1989.

Lucas did not have that repeat X-ray until May 1990, when he was so sick that he had to be admitted to hospital. Further investigations revealed that his right lung had collapsed completely as a result of a mesothelioma. Surgery was performed to remove as much of the tumour as possible and this has given Lucas a short new lease of life. But he knows that his time alive is limited. 'My body is not healed.' He coughs continuously, is losing weight and no longer has any strength. 'People who work with me grumble

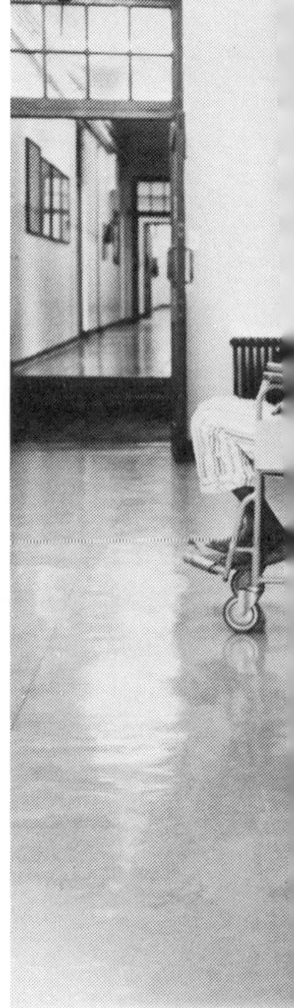
JENNY GORDON

Mesothelioma is an aggressive disease caused by exposure to asbestos. Affected people seldom live longer than fourteen months after diagnosis

because I no longer work like before. The bosses shield me. I am deteriorating every day and more than anything I want a pension — instead I have to work.'

When I first met Lucas in June 1990, he was an angry man who was demanding more recognition from the mining industry. He did not want to touch the compensation money as it was so inadequate. But after a vain search for fair compensation, he resigned himself to this pitiful payout. 'It is useless to be angry. There is nothing I can do. There were many things I planned to do but it does not matter if I'm not successful — my only worry is my children.'

While the Occupational Diseases in Mines and Works Act makes provision for grants for the education of children of incapacitated miners, the racially-based clauses in the act exclude black miners and their families from these benefits (Occupational Diseases in Mines and Works Act, 1973).

Mafefe: 'the dumps of death'

It is because of the suffering of people like Lucas that the asbestos dumps in Mafefe have become known as 'the dumps of death'. The people of Mafefe are concerned about their own health but worry much more about the health of their children. They long for a Mafefe without asbestos.

Mafefe is a district in the north-eastern Transvaal situated in a valley where the foothills of the Drakensberg and the Strydpoort Mountains converge. The district is divided into two sections by the Mohlaphitse River, which flows into the Olifants River, the southern boundary of Mafefe. It is a remote community: the nearest town, Lebowakgomo, is one and a half hours away, while Pietersburg, the nearest city, is two hours distant. A demographic study undertaken in November 1987 revealed that the Mafefe community included 11 202 persons living in 27 villages (Felix et al, unpublished report).

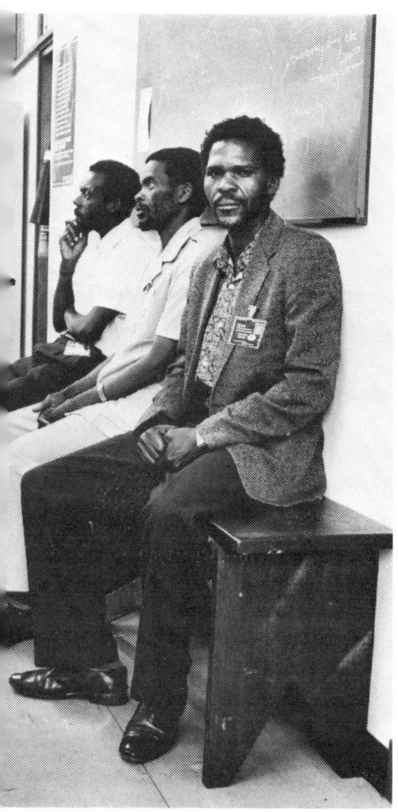

ABOVE
Lucas Maalebogo Maenetja waiting to see the doctor at the National Centre for Occupational Health Clinic, Johannesburg

Mining in Mafefe dates back to the second decade of this century. Numerous small mines developed. From oral histories it emerges that the mining was performed by blasting adits (horizontal tunnels) into the mountainsides where the asbestos seams, contained within the banded ironstone host rock, came to the surface. The ore was brought to the entrance of the adits, where groups of workers would dislodge the asbestos from the banded ironstone with 'five pound' hammers, a process called cobbing. Cobbed asbestos ore was transported by donkeys down the mountainside. Initially the asbestos was crushed and sorted by hand but later mechanized mills were introduced, situated in the valleys close to the roads. The crushed asbestos fibres were packed into jute bags and transported by trucks to depots in Pietersburg. The waste from the milling process was dumped in close proximity to the mills, creating tailings dumps, the major source of environmental asbestos pollution. Even though the mining of asbestos stopped in about 1975, pollution from the tailings dumps continues to this day.

A senior official from the Government Mining Engineer's Office said, 'I didn't even know Mafefe existed'

Nineteen asbestos tailings dumps are located among the villages of Mafefe. Seven border on the sand roads which are in daily use by vehicles, causing fibres to become airborne. Nine of the dumps extend onto the banks of the Olifants River or its tributaries, which are the only water source for the nearby villages of Mafefe. Drinking the water is probably not a health hazard, but washing clothes in the water results in fibres adhering to the garments. Simply by wearing these clothes, people create their own fibre cloud. In addition, there is no closed waterborne sewerage in Mafefe. Any water used is thrown out into the courtyards of the homesteads. Once the water has evaporated, the fibres easily become airborne.

During the early years of mining some of the main roads through Mafefe were surfaced with asbestos tailings. A careful observer will notice that the soil alongside the roads is still finely flecked with asbestos fibres. Asbestos fibres are carried in the dust clouds created by each passing vehicle.

The dumps were used as short cuts to and from schools, shops and rivers. In the rocky terrain, the fine-textured dumps stood out as ideal playgrounds for children after school. Until 1984 children gambolled about in these giant sand pits, in ignorance risking their lives.

Domestic animals graze freely during the day but are kept in kraals adjacent to the homesteads at night. Donkeys and goats often rest on the dumps, the smooth tailings providing a respite from the hot, stony terrain. At night these animals disperse the asbestos fibres that adhere to their hides during the day on the ground around the homesteads.

Asbestos tailings were used by the community as material for building. Women mixed asbestos tailings with mud to plaster the walls of their homes or the floors of their 'lapas' (courtyards) — they admired the blue sheen asbestos gave to their homes. Asbestos tailings were a preferred alternative to river sand in cement brick making and three brickyards in Mafefe all used asbestos tailings in this way. Many residents also made their own bricks.

Asbestos pollution became ubiquitous. For example, in 1963 the Mafefe Primary school arranged for a load of asbestos tailings to be dumped in the schoolyard just in front of the area where pit latrines were to be built. Bricks were made on site but a large pile of asbestos tailings was left unused and, with time, these have been ground into the earth by children walking to and fro to use the toilets. Another dump left in the yard of a local shopkeeper, Mr Seroka, was dispersed by the children who played on it. To this day the soil in the yard provides evidence of asbestos pollution when examined closely. It is heavily flecked with asbestos fibres.

The 1987 demographic study revealed that 35 per cent of the houses in Mafefe were either plastered with asbestos tailings or made of cement bricks containing tailings. As many as 50 per cent of the public buildings were constructed using bricks containing asbestos tailings (Felix et al, proceedings of IUAPPA Conference).

The environmental risk in Mafefe received widespread media

coverage in 1984. For the first time an awareness was raised in the community of the health risks associated with asbestos exposure. As a direct result of this media coverage, residents now seldom make use of the asbestos tailings for home construction purposes and have avoided walking across or playing on the tailings dumps.

The health implications of asbestos in Mafefe

The Mafefe community was anxious to learn more about asbestos. In 1987 the community facilitated a National Centre for Occupational Health (NCOH) initiative to identify the health implications of asbestos fibres in the environment. The NCOH saw its responsibility as threefold:

- to facilitate education of the community
- to initiate research into the prevalence of asbestos-related lung disease and ambient asbestos fibre levels
- to motivate for the appropriate reclamation of asbestos-contaminated areas.

The research philosophy emphasized community participation. This ensured that a group of committed community members came to view the abatement of the asbestos problem and the monitoring of fibre levels in the long term as their responsibility. To prevent unnecessary exposure to asbestos, it was vital that the community understood the hazards of asbestos. Since amphibole asbestos is, for practical purposes, indestructible, it will remain a problem in Mafefe for decades, if not centuries.

The research process allowed fieldworkers to acquire both a comprehensive understanding of asbestos-related disease and the confidence to do community work. Thanks to the determination of Zachariah Mabiletja, employed as a community worker, the training of the fieldworkers was successful. The Mafefe Asbestos Health Workers Committee was formed — an enthusiastic group of twelve committed to ongoing health education work in Mafefe. Their work continues today. The hazards of asbestos contamination are not easy to communicate, but the health workers have composed songs and a play to make their message more accessible.

The research expanded into a number of studies. The first was the demographic study of 1987, the second a medical survey in 1988 of adults who had spent most of their lives in Mafefe. The third involved atmospheric sampling to assess the levels of asbestos fibres in the environment.

The 1988 medical survey was a major event in Mafefe. The demographic survey the previous year had aroused people's curiosity. Everyone knew that a small group of community members was being trained to assist the NCOH research team in a health survey of Mafefe. When people were asked to participate, most were very keen to be involved. A random sample of 833 individuals older than nineteen years, stratified by sex and age, was drawn. A total of 611 people responded, giving a response rate of 73 per cent. All were interviewed and had chest X-rays taken.

Asbestos affects the lungs, causing distinctive X-ray changes.

The children gambolled about on the asbestos dumps, in ignorance risking their lives

PHILLIP VAN NIEKERK

ABOVE

Risking their lives in ignorance: children using the Mahlatjane asbestos tailings dump as a shortcut home from school

Localized thickening (plaques) and diffuse thickening of the lining or pleura of the lungs are the most common diseases caused by asbestos exposure. Recent research suggests that pleural plaques cause lung function impairment (Bourbeau et al, 1990:837-842). A second condition is asbestosis, a scarring of lung tissue, which causes shortness of breath and a persistent cough. Asbestosis usually occurs after considerable exposure to asbestos. Thirdly, an elevated incidence of lung cancer is also associated with asbestos exposure. Finally, a fatal cancer, mesothelioma, which affects the lining of the lung or abdomen, has been associated with minimal exposure to asbestos. All asbestos-related diseases have a long latency period, not manifesting for twenty years or more after initial exposure.

Three experienced readers read the X-rays for the presence of these asbestos-related diseases. All results reflect agreement by at least two readers on the presence of disease. The prevalence of pleural changes for the total sample (611) was 40 per cent (245), the prevalence amongst males being 50 per cent (108) and females 35 per cent (137). For those with environmental and occupational exposure the prevalence was 49 per cent (109), while for those with only environmental exposure the prevalence was 34 per cent (134). Occupation was not known for two participants who had pleural changes. As expected, the prevalence of asbestos-related X-ray changes increased with age. Significant increases existed for both the environmentally-exposed and the occupationally-exposed groups as shown in the diagram on page 43.

The identification of individuals with the less common diseases such as asbestosis, lung cancer, or mesothelioma was not expected. Therefore the finding of a prevalence of 2 per cent (12) of the sample with X-ray changes suggestive of asbestosis was of great concern. Another participant was referred to hospital for investigation of possible pulmonary tuberculosis and was in fact found to have lung cancer.

Tuberculosis is the only other disease that may scar the pleura. But in the sample there was a very low prevalence of active tuberculosis: sputum culture and serum testing revealed only 1 per cent (5) of participants to have active tuberculosis.

To summarize, the effect of the asbestos in the Mafefe environment was shown to be overwhelming, with the prevalence of asbestos-related diseases being much higher than expected.

An important consideration for the community is how scarred lungs affect job opportunities. In applications for jobs where pre-employment examinations are necessary, such as on the mines and in many industrial jobs, people from Mafefe may be at a disadvantage. If their X-rays reveal the presence of pleural plaques, they may not be employed. This is a great worry to the job seekers as their chances of gaining employment are already slim, coming as they do from a rural area and having no ready accommodation in the cities.

It should be remembered that the diseases identified in this study had their beginnings more than twenty years ago, as there is a long latency period between exposure and manifestation of disease. Therefore an environmental atmospheric sampling survey was carried out to establish the current risk. In 1988, 296 samples were collected over a six-month period. An important finding of the survey was a mean fibre count of 10 302 f/m^3 for the 213 samples analysed by light microscopy (Rendall et al). Historical environmental sampling data exists, but careful analysis of this data is necessary before comparisons can be made with the data from the 1988 survey. However, it seems probable that asbestos pollution in the villages during mining operations was not substantially different from conditions today. This implies that the prevalence of asbestos-related diseases may not be any less in twenty years' time. To allow these conditions to persist when a whole community is at risk of developing incurable diseases is unacceptable. For this reason the NCOH has argued strongly for the reclamation of the Mafefe environment.

Reclamation of the dumps is of prime importance. The state has accepted responsibility for the dumps and the Research Institute for Reclamation Ecology based at Potchefstroom University started the cleaning-up process in Mafefe in November 1990. But it is vital that more than just the dumps are reclaimed. Because the asbestos was never regarded as hazardous waste, many of the roads, playing fields, public buildings, and homes are extensively contaminated with it. Therefore reclamation must include the burying of contaminated soil from the roads and sportsfields. Subsidized paint and cement must be made available for the restoration of the homes and public buildings plastered or built with asbestos.

The community response

The community was at first sceptical as no-one in the past had warned them of the dangers of asbestos. Community meetings were held in 1987 to discuss the NCOH's planned work. At one such meeting, when the Moshate (community hall) was jam-packed, an old man sprang to the floor demanding to be heard:

Women mixed asbestos tailings with mud to plaster the walls of their homes or the floors of the courtyards

'Why should we believe in a young white woman when the men we worked with never warned us against it?'

Many people became curious and attended every community meeting to find out more. Some community leaders, teachers and shop owners gained a good understanding of the asbestos problem and were very concerned about the welfare of their community: their watchword became, 'Make Mafefe safe'. But most people in Mafefe continued with everyday life and it is only thanks to the perseverance of the fieldworkers that most households came to understand the hazards of asbestos.

The group of fieldworkers has shrunk considerably. Many used the meagre amounts they earned from the project to further their studies or to seek employment elsewhere. The present group of twelve has been together for a year. Committed to staying in Mafefe and removing the asbestos hazards, they realize that their efforts are for the children of tomorrow. In case they should forget their task they remind themselves in song:

Ruri, ruri,
Borela bo kotsi.
Are bo lweseng, are bo lweseng,
Re bo fenyeng.
Surely, surely,
Asbes is dangerous.
Let's fight, let's fight,
Let's fight 'til it disappears.

If it were not for this group, asbestos pollution would not retain the priority it deserves, since rural people tend to be caught up in urgent, short-term problems. Life is a constant struggle in Mafefe: unemployment is high, money is scarce and any resources must be carefully used. The community have accepted that asbestos is a dangerous substance, but are loath to use their scant resources to minimize the asbestos hazard in their vicinity. For people who have asbestos plastered onto their mud hut walls it is not a problem as they have plastered over the asbestos with a mud mixture. For those who have asbestos bricks, cement and paint are necessary. They are hoping that the state will subsidize the paint and cement so that they can tackle the asbestos problem in their homes.

It seems that the state is not making an attempt to hold those who profited from the mines responsible for contributing to the costs of reclamation work

The state's response

The Atmospheric Pollution Prevention Act of 1965 is a key piece of legislation which should have been applied to protect the environment of Mafefe. The Act stipulates that either the owner of the mine or the state are responsible in a dust control area to 'take the prescribed steps or (where no steps have been prescribed) adopt the best practicable means for preventing such from becoming so dispersed.' For operations after 1965, 'the owner of any mine who ... disposes of any asset of that mine before he has been furnished with a certificate by the chief officer to the effect that the necessary steps have been taken ... to prevent the pollution of the atmosphere by dust ... shall be guilty of an offence.'

If the mine closed down before 1965 and '... if the person liable ... is deceased or has ceased to exist ... the Minister [of Health] may ... cause such steps to be taken ... and direct that the cost involved ... shall be paid by the state, any such local authorities and the owner concerned to such an extent or in such proportions as may be determined by the Minister' (Atmospheric Pollution Prevention Act, 1965).

LESLEY LAWSON

ABOVE
Mafefe asbestos health workers enact their play

The crucial element of the Act is that the minister must declare an area a dust control area, a decision which is often based on advice from the office of the Government Mining Engineer (GME). The task of preventing pollution in dust control areas is delegated to the office of the GME. Before the publicity about Mafefe broke in 1984, no attempt was made to control environmental pollution there. A senior official from the GME's office said at the time, 'I didn't even know Mafefe existed.' This is a serious oversight as the GME's office has records from the early 1930s detailing mining activity in Mafefe. Thus its officials should have been familiar with any previous or current mining venture that posed a health risk for surrounding communities.

In the face of the intensive media coverage the state's response was to start monitoring the fibre levels in Mafefe. Since 1985 the GME has collected eight-hour atmospheric samples at three primary schools in Mafefe on one day every six months. In 1985 the GME commissioned the Research Institute for Reclamation Ecology to cover the asbestos dumps of Bewaarkloof, a valley 40 kilometres east of Mafefe. It is not clear why the reclamation started in the sparsely-inhabited Bewaarkloof instead of in densely-populated Mafefe, but it may have been linked to the fact that Bewaarkloof is in South Africa and Mafefe is part of a 'homeland', namely Lebowa. Another contributing factor might have been that the responsibility for declaring Mafefe a dust control area was shunted to the Lebowa Ministry of Health, who gazetted it only in 1989.

In 1989 the Research Institute for Reclamation Ecology motivated postponing reclamation work in Bewaarkloof and using the limited funds still available for reclamation work in Mafefe. The GME supported their proposal. After extensive surveying of the dumps, earth-moving equipment rolled into Mafefe in September 1990, to the delight of the whole community. But within two months of work starting in Mafefe the Institute's contract was curtailed as the available funds were exhausted. While the work of the reclamation institute is of a high standard it is constantly hampered by a shortage of funds.

Because of ignorance, too few people have treated asbestos with the respect it deserves

> **When a disease is preventable it is hard to believe that the evidence of an epidemic has been ignored**

Even though some of the previous mine owners are still around, the state has taken on the major responsibility for funding the reclamation work. It seems the state is not making an attempt to hold those who profited from the mines responsible for contributing to the costs of reclamation work. The owner of Dalton Mine made a half-hearted attempt at reclamation in 1986, yet the GME has not requested that he assist financially with more adequate reclamation. The attempt took the form of covering the dump with a thin layer of soil and spreading thorn tree branches around the edge of the dump in an ineffectual attempt to dissuade cattle from crossing it. Within months the asbestos was once more visible and being dispersed.

The state's approach does not inspire confidence in the way it is handling a very real threat to the people of Mafefe. The battle against asbestos pollution is being given a low priority. Apart from the human suffering that will result from this policy, it is short-sighted in economic terms as comprehensive reclamation now will substantially diminish the future cost of medical care for the victims. The lack of an integrated state approach to the problem is illustrated by the quality of the health services provided to Mafefe. A clinic exists in Mafefe, but the staff have never developed a service to diagnose, compensate where possible, and care for people with asbestos-related diseases. Even after the media publicity of 1984, no initiative was made to upgrade the health services at the clinic, or ensure that appropriate health services were provided to deal with the asbestos-related epidemic. All people with chest diseases have to travel to Groothoek Hospital, 100 kilometres away by poor roads. The trip is costly so many go to hospital only when the diseases they are suffering from have reached an advanced stage. In addition, many are misdiagnosed as tuberculosis sufferers and receive inappropriate medical treatment.

Mine owner responses

Msauli and the Griqualand Exploration Financing Company (GEFCO) are the two large asbestos mining companies that own most asbestos mining claims. Both are still actively mining asbestos. Clearly GEFCO was concerned about the negative publicity the mining companies received during the media coverage in 1984. As a result, it has taken the issue of hazardous waste seriously and has commissioned the Research Institute for Reclamation Ecology to reclaim the dumps and clean up areas surrounding their operational mines. Neither appears to own any mining rights in Mafefe, but GEFCO owns the mining rights to Egnep, one of the mines in Bewaarkloof. GEFCO was expected to contribute a mere 5 per cent of the cost of reclaiming Egnep, with state and local authorities paying the rest. GEFCO paid slightly more than the required amount. This 'generosity' gained GEFCO credibility among state officials, although it could be argued that GEFCO should have paid the full costs involved according to the 'polluter pays' principle. Gencor, the previous owner of GEFCO, sold GEFCO to its directors in 1989. This is thought to be in response to the possible threat of litigation.

ENVIRONMENTAL vs OCCUPATIONAL EXPOSURE

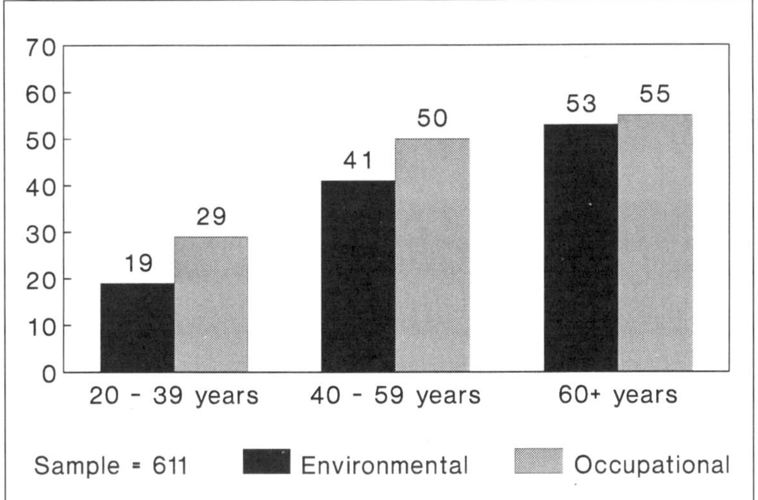

A graph showing the prevalence of changes of the lung lining according to age and environmental or occupational exposure

Mafefe is not alone

People in Kuruman, Prieska, Pomfret, Barberton, and Badplaas share a similar plight to the community of Mafefe. The list of places polluted with asbestos is depressingly long. South Africa is the only country in the world where all three commercial types of asbestos have been mined. Crocidolite and amosite are found in the Pietersburg Asbestos Fields, where Mafefe is situated; crocidolite is mined in the north-western Cape; and chrysotile is found in the eastern Transvaal. Mining still occurs in all three regions.

However, asbestos pollution is not confined to mining communities. Most of the asbestos is exported, so dockworkers and railway workers are exposed, as well as populations in other countries. Local industry has manufactured insulation materials, brake pads and cement pipes, to name but a few products. During the production process workers are exposed, sometimes to very high fibre levels; and users of these products have never been properly informed about how to minimize exposure during use. As a result of their ignorance, too few people have treated asbestos with the respect it deserves.

The consequences have been painful and tragic for thousands of families who have witnessed the different diseases take hold of their loved ones. Research shows that South Africa has the highest rate of mesothelioma in the world (Zwi et al, 1989:320-329). Preliminary results of a birth cohort study of 399 whites born in Prieska from 1932 to 1936 reveal that of the sixty-six people who died, six died from mesothelioma, a staggering 9 per cent (Reid et al, 1990:584-565).

When a disease is preventable it is hard to believe that the evidence of an epidemic has been ignored for so long. But as more people realize the risk imposed on them by environmental conditions that the state, the asbestos mine owners, and medical professionals have allowed to develop, the public outcry will grow into one that cannot be ignored.

■ Lucas Maalebogo Maenetja died on 30 May 1991 of mesothelioma.

•TSHEPO KHUMBANE•

If somebody else hadn't said that to give a man a fish is to feed him for a day, but to teach him to fish is to feed him for life, Tshepo Khumbane would have. It is an integral part of a philosophy that has grown and developed through a life spent in the service of others.

Teacher turned social worker turned field worker, Khumbane now works with the Environmental Development Agency (EDA). She has, since her first post — in a mission hospital at Hammanskraal — been aware that handouts are never a satisfactory answer and that even helping individuals is not enough. Real help means reaching out into the community and involving them in self-help projects.

Khumbane came to that conclusion the hard way in Hammanskraal. There was no back-up, no welfare structure, no follow-up procedure. In that poverty-stricken, malnourished community, many of the patients suffered from tuberculosis — a problem, Khumbane realized, which affected the whole family.

'So what do you do when you get to an area where there is absolutely nothing? At home the wife is sickly, there is no food, no employment. I felt it was a useless venture to be calling people and interviewing them and filling pages while the situation remained as it was.'

So, she called together the sisters, mothers and wives of hospitalized TB patients and began to look at their problems. A major difficulty was helping children stay at school when no money was coming in and they couldn't afford to buy books. The solution was to collect books through the church and distribute them.

In 1966, Khumbane moved to Lesotho with her husband and her work with communities continued. Through her involvement with patients in a leprosy asylum, she quickly realized that, here again, the problem lay with the community not merely with those afflicted with the disease.

Funds were raised for an orphanage for children whose parents could not take care of them and Khumbane and her team began to work to integrate the patients back into their own community — by bringing the community into the asylum. 'We had to open up the barrier between them.'

A village home-workers' scheme through which the disabled were taught crafts was extended to the asylum. Funds were raised, machines and materials were donated and the patients began to make their own clothes.

'It helped to reconstruct those people,' says Khumbane. 'They came to realize they were people with a role'. In the process, Khumbane realized something too. 'In Lesotho for the first time, I understood what was going on. I felt a very strong push that it wasn't welfare work that could achieve. I became more development-orientated.'

At home she gardened, planted fruit trees, raised chickens, knitted and sewed. It was time, she thought, that she taught others to do the same. 'We had to try to make people think and face their problems and deal with them.' With a group of women she formed a small sewing co-op. With the vagrant youths who had come to Maseru looking for work she built temporary shelters with mud bricks and around them planted vegetables.

One particular programme organized by the Lesotho government impressed Khumbane greatly. One day a year was set aside as tree-planting day, with every person in the kingdom being expected to participate. It was a concept she took back with her when she returned to South Africa and to the Hammanskraal mission hospital at which her career had started.

There she organized health days centring on the value of 'growing your own'. Seed was collected from neighbouring white farmers and distributed to patients at the clinic and to those reached by mobile clinics. 'It made more sense than packing food rations in the hospital. We could say, "this is permanent".'

With the cooperation of health officials, clinic nurses, agricultural officers and church ministers, a nutrition committee was formed. Made up of women in the villages, its chief function to educate. By the time Khumbane left Hammanskraal, she was firmly set on the development path. Moving to an EDA programme in Bochum in the northern Transvaal, she discovered even worse poverty than she had found before — and nobody seemed in the least interested in doing anything about it. 'I was placed at the magistrate's offices with no facilities. If I hadn't had the imagination and drive I could have ended up being a clerk registering births and deaths.'

LESLEY LAWSON

Instead she started a women's group. 'I was worried about the nutritional level in the area and the unemployment. We needed to build confidence in the women to face poverty and understand that it can be relieved. I closed that office for two weeks and went into the villages, started groups, planted vegetables. I also saw tree-planting as something that could mobilize the whole district.'

Within a year, the district was indeed mobilized. It would hardly have dared not to be — Tshepo Khumbane is a very determined woman. There were 1 500 women in thirty-one villages involved in working out programmes of education and agricultural action. Sometimes they would walk for 17 kilometres, starting at three in the morning, to reach far-flung villages. They took lessons at the agricultural offices and sold the results of their labours.

An umbrella organization was formed by the women to coordinate efforts. School children and teachers were brought in. A nursery was started. For five years, says Khumbane, 'we planted trees and trees' in an attempt to make the land live again.

Sadly, though, in that hot and sandy place where water is always a problem, few of those trees survived. 'We tried to tackle the environmental issue, we involved almost everybody in the community, but we didn't look at how they could maintain trees without water. There was a vision but we didn't look into where do we start.'

It has been disappointing, but Khumbane is not deterred. Full of positive enthusiasm and energy ('at times you feel you can just wake up and keep planning and doing'), she still sees tree-planting as a priority if the erosion problem is to be controlled. She believes there must be education about a wood-lot. First, though, it will be necessary to tackle the water problem. Only then, says Khumbane, can the community plan to tackle environmental issues for the future; only then can they hope to 'dress the land'.

Khumbane is aware that the programme will require a great deal of cooperation between the communities involved, their chiefs and local government, and is under no illusions about the very real problems of internal disputes on the one hand and the absence of water on the other.

'But I don't think it's so hopeless nothing can be done. We only want the chief to say yes. With permission we will mobilize the whole community. And once we have that water, there will be a green carpet.' ❐

4 THE GHETTO AND THE GREENBELT

The environmental crisis in the urban areas

◆ LESLEY LAWSON ◆

Norweto and the garden village

'Is it the security, the landscaping or the view that has made Fourways Gardens the fastest-selling suburb in Sandton?' (Advertisement in *The Star*, October 11 1986.)

In 1986 *The Star* Homes Show treated the public to a new concept in suburban living. The Fourways Gardens Village — which comprised 300 stands, a nature sanctuary, a trim park and an adventure playground — was completely walled, its single entrance guarded day and night.

During the seven days of the show visitors streamed in from all over the city. The only blot on the landscape was a pamphlet protest by members of the Greenbelt Action Group (GAG). They claimed that the promoters were deceiving the public by failing to inform them that this new paradise was situated 5 kilometres from the proposed black township of Norweto.

Angry words were exchanged. The property developer told *The Star* (22 October 1986) that as there were two valleys, a golf course and a major highway between the two sites 'there is no way that the view from Fourways Gardens would be affected by Norweto'.

GAG's objections to Norweto were largely that it would destroy the ecology of the area, causing pollution and social disruption. Its members were adamant that 'the political climate in the country has changed and there is no longer room for black ghettos in greenbelt areas' (*The Sunday Times*, 19 October 1986). GAG's opposition was strengthened by support from the then Progressive Federal Party and the Conservative Party. The former complained that Norweto was racially structured and economically unviable, while the latter adopted a more straightforward swart gevaar position.

Various spokespeople from the black community rejected Norweto on the grounds that it was too far from employment oppor-

'It sometimes appears as if it is easier to raise R2 million to save a rhinoceros than to save a human from a life of deprivation'

tunities — and they renewed their demands for more affordable, accessible housing. The government quickly dropped its Norweto idea — but failed to propose a more suitable alternative site for low cost housing.

The Norweto controversy provided an insight into contradictory views on the concept of environment in the different sectors of our population. For many white environmentally-conscious people, a responsible attitude to the environment means preserving access to greenbelts and recreational areas for urban dwellers. It also means preserving low-density residential areas and well-serviced roads and parks.

For black South Africans, a healthy environment is what is required for healthy living. In the words of Japhta Lekgetho of the Soweto-based National Environmental Awareness Campaign:

> It sometimes appears as if it is easier to raise R2 million to save a rhinoceros than to save a human from a life of deprivation. Surely nature conservation is also about the preservation of people? (*The Daily Mail*, 24 July 1990).

For many white environmentally-conscious people, a responsible attitude to the environment means preserving greenbelts, low-density residential areas and well-serviced roads and parks

Urbanization and the environment

Throughout this century, a central concern of the ruling white government has been to prevent black urbanization. The legacy of this policy — segregated, degraded, overcrowded black urban areas — is the core of the urban environmental crisis in South Africa today.

The absence of housing and basic infrastructure, and the existence of influx control and pass laws until 1986, did not deter the urbanization process. However, the chronic lack of space and facilities for low-income people in the urban areas has created huge problems that will have to be solved in the future.

South Africa's urban population continues to grow with unprecedented speed. Anglo American's Clem Sunter claims that the informal settlements in the Durban area grew at the same rate as those of Mexico City in the 1980s (Huntley et al, 1990). The Urban Foundation's research (Urban Foundation, July 1990) suggests that some 750 000 black South Africans are moving into the cities every year, and that since 1988 there have been some 7 million people living in shacks and other informal housing — half of these in the PWV area.

Any discussion of urban environmental problems in South Africa must be seen against the backdrop of this ongoing process of urbanization. The overcrowding of existing townships is putting the already inadequate housing and service infrastructure under impossible strain. People who establish their own informal housing on vacant ground suffer under even worse conditions — a lack of access to safe water, proper sewerage and refuse removal.

The inadequacy of the government's response to the needs of the growing urban population is having consequences both on the health of the people and on the long-term integrity of the environment. Although its particular history sets it slightly apart from other townships, Alexandra township provides a potent case study of our urban environmental crisis.

Alexandra township

'Approaching Alexandra township by air shocks the senses more than it would when coming in on the ground. One moment you are gliding over large beautiful homes set in lush gardens amid tennis courts, stables and sparkling swimming pools, then the earth below you suddenly turns brown and scabrous as if it had died,' said David Braun in *The Star* on 26 July 1986.

Alexandra township — usually referred to simply as Alex — is 2 square kilometres of black residential space bordered by the white Johannesburg and Sandton suburbs of Wynberg, Lombardy, Kew and Marlboro. It was established as a black freehold area at the turn of the century.

A thorn in the side of the Nationalist government's separate development policy, Alexandra was scheduled for reconstruction as a migrant-labour hostel city in the 1960s. In line with this decision, much of the existing housing was destroyed and thousands of people moved into Soweto and other distant townships.

For the next twenty years Alexandra was regarded as a temporary residential township, and there was little expenditure on housing or services like water, refuse removal, electricity, sewerage systems, roads or stormwater drains.

In 1979, as part of the government's new reform policies, Alexandra township was reprieved. By this stage the township environment had become seriously degraded (Alexandra Liaison Committee, 1983). A lack of stormwater drainage had reduced the untarred roads to potholes and dongas, and mountains of refuse lined the streets. Water was provided by outside communal taps — one for every three houses — and the sewerage system consisted of bucket lavatories that were irregularly collected. Few houses were electrified, and pollution from coal stoves created a pall of smoke over the whole area.

LESLEY LAWSON

At this time Alexandra was already considered to be seriously overcrowded — with a population of 75 000 living mainly in backyard shacks.

Since the mass protests and struggles of 1986, the government has spent R138 million on the upgrading of Alexandra — tarring certain roads, installing waterborne sewerage, electricity and stormwater drains. No low-cost housing has been built, but a private sector housing scheme on the East Bank has provided an additional 900 houses for wealthier residents.

Alex currently has a population density of nearly 800 people per hectare

The continuing crisis in Alexandra

Despite the government's upgrading programme, the situation in Alexandra township has deteriorated dramatically. First, given its political and financial weaknesses, the black local authority has been unable to maintain any of the new services. Second, because of its location, there has been a dramatic influx of people to

Alexandra from surrounding townships.

Despite the poor conditions, a house in Alexandra township is a valuable asset. Alexandra is the only black residential area near Johannesburg and Sandton where low-income people can obtain access to the city, with the employment and economic opportunity that this implies. With room rentals at R14 a month, it is also one of the cheapest places to stay.

In the last four years the Alexandra population has soared, and now stands at 175 000. It is continuing to increase at the rate of 300 people a month: this gives a current population density of nearly 800 people per hectare.

ABOVE
The old and the new …
Alexandra township with new housing rising behind the east bank of the Jukskei River

The Alexandra Civic Organization (ACO) has taken on the challenge of the housing crisis. In 1989 they contracted Planact, a service organization involved in urban planning, to study the housing and service situation in Alexandra. Planact's Andrew Boraine has provided a detailed description of their findings. He states that Alexandra has one of the highest ratios of shacks to houses out of all the townships he has seen. In the old part of Alexandra, at least 70 per cent of the residents are in informal housing. The breakdown is as follows:

- 13 962 families in single rooms attached to houses
- 11 000 families in freestanding shacks
- 6 120 families in backyard shacks
- 379 council houses
- 1 452 flats
- 1 662 private houses
- 8 432 hostel beds

'The piles of refuse lying around create a sense of helplessness in the people'
— Richard Mdakane

The services installed during the upgrading plan have failed to cope with the growing demand.

An example of this is the system of stormwater drains — which functioned properly for only one season before deteriorating through lack of maintenance. This has had a marked effect on the quality of the roads. The few roads which are tarred have begun to crack, while the majority of the roads which are not tarred are scoured into deep dongas by the summer rains. Motorists who are strangers to Alexandra have to be guided around the township during the rainy season, as many of the roads are impassable.

During the upgrading programme over 4 000 communal washing units and toilets were installed — on average one per yard. Boraine estimates that each unit is shared by an average of fifteen families — sometimes even more when there are informal settlements on adjacent vacant lots. This shortage of facilities, especially regarding water, is the source of ongoing conflict. Tensions break out in the yards, and sometimes landlords place locks on the water taps to prevent their use.

The communal units frequently break down from overuse, and are not properly maintained. There are reports of toilets staying blocked for months, with residents being obliged to use the streets as alternative lavatories.

Parts of Alexandra do not have waterborne sewerage at all and still use the bucket system. The buckets are supposed to be collected every week, but this system too is irregular. It is common to see rows of overflowing buckets next to the toilets, waiting to be collected.

Apart from the dangers to human health, this unhygienic system also pollutes the Jukskei river, which flows 10 metres away from the site where the buckets are cleaned.

The inadequacy of the refuse removal system further degrades this township. A private company is contracted to remove the household rubbish, but he has the capacity to clear only a third of Alexandra at a time. The Council has the responsibility of cleaning the streets, but also lacks the means to do so.

Some streets in Alexandra are bordered by garbage piles three metres high. Children play among the carcasses of dead dogs and piles of rusty cans. Cows graze peacefully from dustbins overflowing with rotting vegetable matter.

Pollution

Alexandra township is on a hill leading down to the Jukskei River, which receives a considerable amount of stormwater runoff. The pollution level of this water has not been measured, but is thought to be fairly high. People cross this river on foot every day and, owing to the shortage of playgrounds, it is the children's favourite haunt.

The programme to electrify Alexandra has proceeded slowly and at this point less than a quarter of the township has electricity. The majority of residents have to rely on coal for energy, which causes health-threatening air pollution.

The situation is even worse for the shack dwellers who must rely on braziers for heating and cooking. These are an even dirtier source of energy than coal stoves, and are frequently used inside rooms where there are no chimneys.

Affordable houses for all

ACO has embarked on a vigorous campaign to improve housing in particular and the township environment in general. ACO argues that there is a direct connection between finding solutions to the housing needs of thousands of low-income families, and the preservation of the environment.

With Planact, they have drawn up plans to develop an area of land on the east bank of the Jukskei River, into a low cost housing project. They propose that the project be owned by a Land Trust and run democratically by the Alexandra community. They say that only in this way can they avoid the problem of the private sector building houses which most Alexandra residents cannot afford. ACO submitted a tender for this project to the Transvaal Provincial Administration (TPA) in July 1990, along with several private

LESLEY LAWSON

LEFT
Rotting debris and garbage on the banks of the Jukskei River, which runs through Alexandra township

developers, and is awaiting the outcome of an adjudication process.

A pamphlet from ACO to the community ('Housing for all in Alexandra', 1990) explains their intentions:

> If the people of Alexandra are going to have a direct say in the development of the Far East Bank, then democratic structures must be built throughout Alexandra. There are many different people in Alexandra, all with different needs. These different needs should be discussed within the structures.

ACO is also intending to use the project as an opportunity to give Alex residents skills training. They are planning training programmes which will teach people how to 'manage projects, control money and understand technology'.

The Land Trust will also be a major employer in the township and will involve itself in negotiations for extra land.

Local government

ACO has begun negotiating with the TPA and the Sandton City Council for a new form of local government for the township. They have called for the creation of a single non-racial form of local government covering Alexandra and the surrounding white municipalities, resting on a common tax base.

In the short term, ACO has requested that the Sandton City Council take over the refuse removal system — but this suggestion has not been taken up.

Richard Mdakane of ACO says, 'In principle they agree that Alex and Sandton should be one city. But only as long as it does not affect their pockets. The Sandton City Council has the labour force and the trucks for garbage collection but Sandton is hesitant to help us' (Personal interview, 20 December 1990).

Environmental campaigns

Mdakane says that he regards buildings as an important part of the environment. 'You cannot divorce housing from the creation of a liveable environment. But in Alex we need to go beyond the provision of housing. We also need to create facilities like parks, creches and schools to make it a liveable environment. There is not a single park in Alexandra.'

ACO regards the refuse situation as a major problem for Alexandra. Says Mdakane, 'The piles of refuse lying around create a sense of helplessness in the people.' Over the last year, ACO has organized three successful clean-up campaigns. Trucks were hired and members of the community worked together to remove the refuse from the streets.

ACO also began a tree-planting campaign, but this was less successful. Trees in certain areas were uprooted by angry residents who said they needed houses, not plants. Others complained that they would have preferred to have fruit trees.

Forty-four gallon drums have also been collected with a view to establishing a recycling campaign for paper, cans and bottles.

Mdakane says that ACO would like to start a major environmental campaign in the township — but that the white residents of

'The white residents of Sandton need educating too. The water and air pollution from Alex affect them as well'

— Richard Mdakane

Sandton need educating too. The water and air pollution from Alex affect them as well, and they should see the need for Sandton and other municipalities to take on some of the burden of Alexandra.

The suburbs: long-range destroyers

Travelling back from the degraded streets of Alexandra to the leafy lanes of Fourways Garden Village should bring a sigh of relief to the environmentally-conscious.

But these carefully-tended, suburban areas — and the lifestyle that goes with them — create long-term environmental damage too. This damage includes air pollution resulting from the generation of electricity and from industrial processes, as well as the creation of waste through industrial and domestic activity. Other chapters in this book deal with these environmental problems in more detail, but two points are appropriate here.

The carefully-tended suburbs — and the lifestyle that goes with them — create long-term environmental damage too

First, unlike pollution created by township coal smoke, the pollution created by the profligate consumption of electricity does not settle in the lungs of the suburban consumer. Instead most of it ends up in the eastern Transvaal Highveld, which is the one of the most polluted parts of Africa. (See Chapter 10.)

Second, the vast piles of waste resulting from a consumerist lifestyle do not line the streets of the suburbs. Instead they choke landfill sites around the edges of the cities, threatening the underground water table and the health of black communities who live nearby (*Rotating the Cube*, 1990). It is the townships — many of which are situated on the outskirts of the suburbs near dumps and factories — which frequently bear the brunt of industrial air pollution and toxic waste, even though township residents may derive little benefit, direct or indirect, from the processes that produce them.

Health hazards in the township environment

Alexandra township presents an acute example of the health hazards which threaten township residents. But many other townships around the country are experiencing at least some of these environmental stresses as well.

Medical doctors commissioned by Planact have detailed the effect that these stresses may have on the health of residents of townships and squatter camps (Buch and Doherty, 1989).

- Water-related illnesses are caused either by drinking unsafe water or by economizing on washing owing to a shortage of water. The illnesses include diarrhoea, dysentery, worm infestations and a wide range of skin and eye infections. The doctors note that childhood diarrhoea is the primary killer of children in developing countries. Water shortages also create the risk of outbreaks of infectious diseases like poliomyelitis and typhoid.
- Excreta-related infections are caused by inadequate sanitation systems, such as the bucket system, or by leaking waterborne sewerage. These illnesses include beef and tapeworm infestations as well as diseases carried by flies.

LESLEY LAWSON

RIGHT
Squatter huts surround a municipal council notice warning people of a dangerous catchment area

- Refuse-related diseases result from a shortage of refuse bins and heaps of uncollected refuse. They include diseases carried by flies and rats, as well as rat bites, poisoning and injuries, especially to children.
- Health risks are presented by pools of stagnant water resulting from inadequate drainage. These pools are especially hazardous for small children.
- Diseases of overcrowding occur. Overcrowding encourages the spread of airborne infections, the more serious of which are tuberculosis, pneumonia, and measles. Pneumonia is the second most prevalent cause of child death in the world, while measles is the third.

Broad-ranging studies of environmental health hazards in overcrowded townships and squatter areas have not yet been made, but some of the following examples give an idea of the extent of the problem.

Some specific examples

Botleng

Children play among the carcasses of dead dogs and piles of rusty cans

Dr Erich Buch has assessed these hazards in a 1990 study of the Botleng township near Delmas. The township, which has communal taps in the streets and a bucket sanitation system, is home to 55 000 people — the majority of whom are squatters.

Buch's study concludes that 'the environmental health situation in the housing areas of the township is not satisfactory. In the squatter areas it deteriorates to a situation with major health risks' (Buch, 1990).

Soweto

In a 1984 study on Soweto, Dr Tim Wilson concluded that the infrastructure and services were more or less adequate, but he identified overcrowding as one of the important factors influencing the spread of infectious diseases (Wilson, 1984). He said that in the period between 1970 and 1980 only 5 000 houses were built, while the natural increase in the population was 25 000, and that this was leading to an acute housing crisis. He estimated that there were about twenty people sharing each four-roomed house.

Since then the situation has deteriorated. Estimates based on aerial photographs show that there are 97 000 shack dwellers in Soweto — an increase of 22 per cent in the last two years. Another study has shown that 41 per cent of Soweto's houses have backyard shacks (*The Star*, 11 July 1990).

Not only does overcrowding create favourable conditions for the spread of infectious diseases like tuberculosis, pneumonia and measles, but it also impacts on other essential services. And when water and sewerage systems are threatened the health risks increase.

Erich Buch and Jane Doherty (1989) describe these risks in the context of service disruption during the Soweto rent boycott. They state that the water supply was frequently cut for prolonged periods. Sewerage pipes were left unsealed and leaking and refuse was seldom removed.

Buch and Doherty conclude that the resultant unhealthy water, poor sanitation and accumulating refuse were a threat to the physical and mental health of the townships residents.

It must be said that it is usually the women of the community who bear the burden of these extra stresses. It is they who must walk long distances to what are hopefully safe supplies of water, and struggle to cook and keep house on limited water supplies.

It is usually the women of the community who bear the burden. It is they who must walk long distances to what are hopefully safe supplies of water, and struggle to cook and keep house on limited water supplies

Wesselton

In the latter part of 1990, Conservative Party-controlled local authorities began responding to township rent boycotts by suspending services altogether. A newpaper reporter (*The Star*, 25 October 1990) quoted an Ermelo doctor as saying that there had been an outbreak of diarrhoea among Wesselton township residents after a nine-day water cut. He said that 60 per cent of his patients in that time had been suffering from diarrhoea. The reporter also discovered people 'drinking water from a murky dam which had a strange smell'. At this dam was a mother using the water to mix powdered baby milk.

Demolitions

An even worse health threat is posed by the authorities' concerted efforts to demolish informal settlements.

The Dobsonville squatters — whose shacks were demolished at least three times during 1990 — were a case in point. This community suffered an outbreak of measles and bronchitis after spending nights huddled together in the open veld (*The Daily Mail*, 7 July 1990).

During another demolition women showed their desperate determination to protect their homes by standing bare-breasted in front of the bulldozers and shouting, 'We want houses!'

Township air pollution

Further threats to the health of township residents are posed by coal smoke pollution from coal stoves and braziers.

> South Africa's level of domestic electrification lags far behind the global average, and slightly behind the average in developing countries (Dingley, 1988).

The approximately 20 million South Africans without electricity have to rely on dirtier sources of energy — like wood, coal and paraffin — for fuel. These create a pall of pollution which hangs over South Africa townships, especially during the winter months.

Even townships which do have electricity suffer from this problem. Backyard shacks and squatter areas are still not electrified and the use of smoky braziers is common. Other residents are unable or unwilling to replace their coal stoves with electric ones, and so these remain in use. A study has shown that coal stoves are used in many Soweto households that have electricity. Up to 11 per cent of houses that have always had electricity use coal stoves and 22 per cent of newly electrified houses still have coal stoves. Factors which influence this are the expense of buying a new appliance, and the fact that coal stoves are also used for heating in winter, as well as for water heating (Heyl, 1988).

Soweto has been the subject of several air pollution studies. An overview by Kemeny et al (1988) has compared smoke concentrations at different Soweto sites between 1972/3 and 1982/3. The conclusions are as follows:

- The smoke concentrations recorded in Soweto are comparable with those observed in the City of London between 1954 and 1964.
- The smoke concentrations of 1982 were higher than those of 1973. This increase in air pollution during a period when the township was being electrified is probably related to the dramatic increase in the number of people living in informal housing and making use of homemade coal stoves or braziers for heating and cooking. There was also a change in the weekly cycle of pollution over the ten-year period, which the researchers attributed to the use of electricity.
- The level of smoke pollution in Soweto frequently exceeded the guidelines put out by the Department of National Health and Population Development. The Department suggests the following smoke limits: for short-term exposure, a twenty-four hour mean concentration of 25 micrograms per cubic metre and for long-term exposure, an annual mean of 100 micrograms per cubic metre. At one Soweto site during 1982, Kemeny recorded a winter monthly mean of 520 micrograms per cubic metre. The highest daily mean

in this period was 1 040 micrograms per cubic metre and the highest hourly concentration was 4 520 micrograms per cubic metre.

The exact effect of this pollution on the health of township residents is the subject of a current major study headed by medical researcher Yasmin von Schirnding. The study — which will be completed only in ten years' time — will examine the effect of pollution on 4 000 schoolchildren in the greater Johannesburg area.

Von Schirnding says that one of the aims of the study is to look at the relationship between township pollution and acute respiratory infections (ARI). These infections are now becoming a major cause of death and disease among children throughout Africa. In South Africa ARI is now becoming a more important cause of death than diarrhoea. Respiratory infections in children constitute the major reason for their use of the health services (Von Schirnding, 1990).

It must be pointed out that although township residents are the ones to suffer most from township coal smoke pollution, the damage is not confined to those areas. In densely populated areas like the PWV, township air pollution must have an impact on the air quality of neighbouring suburbs.

The smoke concentrations recorded in Soweto are comparable with those observed in the City of London between 1954 and 1964

Water pollution

There have been several recent indications that the inadequate water and sanitation infrastructure in the townships and squatter camps is a serious threat to human health.

Durban

The CSIR is currently engaged in a study of the water quality in the greater Durban area. They have compared pollution levels in the run-off from informal settlements, dense formal housing estates, industrial zones, subsistence farming areas and developed residential areas. Runoff from all of these showed significant levels of contamination with faecal bacteria. The highest level — one million coliforms per 100 ml — came from the informal settlements. This result should be seen in the light of recommended standards, which are 200 faecal coliforms per 100 ml for recreational water and zero for drinking water (Simpson, 1990).

Delmas

Information obtained by Planact during an interview with the Delmas Town Clerk, a Mr van Rensburg (22 August 1990), showed that the drinking water of both Botleng township and the white town of Delmas was threatened by contamination with faecal bacteria. This is because the water comes from boreholes — one of which is 300 metres from the squatter camp. The situation is exacerbated by the fact that the land is dolomitic. According to the Town Clerk the Botleng water had an *E Coli* count too high for human consumption for two months, and the township residents had to get their water from Delmas.

LESLEY LAWSON

RIGHT
Viva Park, a squatter camp that developed in the grounds of a hostel with the Sasol factory in the background

Some say that their eyes stream in response to sudden wafts of fumes, while others complain of recurrent headaches, sinusitis, tonsillitis and respiratory infections

Case study: Zamdela township

Zamdela is a typical small township of some 40 000 people located in the Orange Free State. Like Alexandra, Zamdela has seen a recent influx of homeless people from rural areas. Many of these people have been accommodated in converted single-sex hostels.

One of these — Viva Park — is home to 200 families. Sixty live in the hostel building and another 120 have set up shacks and shanties in the open space within the hostel walls.

There are communal toilets and showers for men and women, but since a recent incident of rape, the women have preferred to use buckets at night. There is no refuse removal system and no drainage for the compound.

On a crisp winter's evening the air is thick with smoke from braziers. On a fine summer's morning dead rats and vegetable peels lie on the ground between pools of stagnant water.

What is unusual about Zamdela is that it is situated downwind from Sasolburg — the heart of South Africa's petrochemical industry. Over and above the trials of living in a degraded urban environment, Zamdela residents must endure unrelenting waves of industrial pollution.

Some say that their eyes stream in response to sudden wafts of fumes, while others complain of recurrent headaches, sinusitis, tonsillitis and respiratory infections. One woman who came from QwaQwa said she had a headache for her first two months in the township. Another resident said he was moving to Sebokeng on doctor's orders because the air pollution was exacerbating his daughter's asthma. A local shopkeeper complained that his car roof was corroded. 'If this is what a car looks like, can you imagine what it is doing to human lungs?' he asked.

The factories in the area — Sasol, AECI and Sentrachem, to name just a few — are collectively responsible for a stream of gaseous waste of unknown composition. And their chimneys are close enough to Zamdela for residents to describe the changing thickness and colour of the smoke.

Not only does proximity to the factories threaten the quality of the air, but the storage of inflammable gases on the factory premises gives cause for concern. Visiting trade unionist from the British Manufacturing Science Finance union, Judith Church, alleges that chlorine gas is being stored unsafely on the AECI premises (Personal interview, August 1990). After visiting the site she said there was the potential for a massive explosion and gas leak which could have Bhopal-type consequences.

Indeed, when guerrillas attacked the Sasol plant in 1980, the fire raged for days. Although township residents escaped injury, some remember spending nights in the veld on the far side of the township for fear of explosions.

Zamdela residents have discussed forming an environmental group to approach management about the level of air pollution, but the volatile political situation in the township has prevented this from materializing.

The siting of townships

The case of Zamdela illustrates a common feature of black townships: they are frequently developed on sites that would be unacceptable to a white community. There are many permutations of this problem.

Dumps and dirty industries

Zamdela is on the downwind side of the Sasol industrial complex; the white town of Sasolburg also experiences air pollution, but it is on the upwind side. And the suburbs of Sasolburg are even further away from the factories on the other side of the town.

Other examples of this abound. On the Witwatersrand many of the black townships, such as Riverlea and parts of Soweto, are situated in the shadow of mine dumps and are affected by the dust and other pollution from mining activity.

Unsuitable land

Many Transvaal townships are built on unstable dolomitic ground. Sinkholes can develop and in some parts houses are beginning to crack. This represents a threat to some 2 million township residents. The areas particularly at risk are Khutsong township, Duduza, Vosloorus, Tembisa, Bekkersdal, Atteridgeville and Lenasia (*The Weekly Mail*, 15 June 1990).

The case of dolomitic rock leading to pollution of the water table has already been discussed. The Town Clerk of Delmas has also said that the dolomitic ground on which the squatter camp has grown will prevent them from ever being joined to the main sewerage system (Personal interview).

The instability of the ground also prevents the further expansion of township areas. For example, 40 per cent of the available land to the west of Soweto is dolomitic and unsuitable for development (National Land Committee, 1990).

Distance from employment centre

The common aspect of the siting of townships is the distance from city centres and industrial areas. Travel time adds hours onto the working day and further degrades the quality of life.

The transport infrastructure required to carry large numbers of workers over long distances every day has its own impact on the environment in terms of fuel consumption, air pollution and loss of habitats to roads, etc.

Quality of life

The quality of life in black townships is also badly affected by overcrowding resulting from the housing shortage. In addition, the poor quality of housing (badly built structures; no ceilings, damp-proofing or insulation; asbestos roofs) is a major problem. Finally, there is the lack of open space, grass and trees.

'If this is how corroded my car is, can you imagine what the pollution is doing to human lungs?'

— Zamdela resident

In conducting struggles and campaigns at a grassroots level, civic organizations have become particularly aware of the need for a healthy environment

Healing the urban environment

It is clear from the discussion so far that the crisis in South Africa's urban areas is deeply rooted in the political and economic folly of separate development. But the legacy of apartheid will remain long after the legal system of apartheid is dismantled.

Solutions to the urban crisis should be seen in terms of access to residential land for all urban dwellers, not just those who can currently afford it. In other words, we should avoid replacing racial apartheid with economic apartheid.

It is essential that residential areas for the urban poor be located near places of work and economic opportunity. In addition, basic services such as water, sewerage, electricity and refuse removal, and the provision of suitable housing for all our people, must be high on the agenda.

New forms of non-racial and democratic local and metropolitan government will be essential if these changes are to take place. The central state, with its access to financial and technical resources, will also have to play a leading role.

However, the key to the creation of 'liveable' urban environments lies with the communities themselves. There are already encouraging moves in this direction. For example, civic organizations in all the major urban centres of South Africa are busy preparing themselves for a developmental role in the future. This marks a shift away from the politics of mobilization that tended to characterize the anti-apartheid protests of the 1970s and 1980s.

Civic organizations have made progress in exploring appropriate forms of organization for community development. These include community-controlled land and housing trusts, credit unions and community loan funds, and community development corporations.

Civic organizations have also begun to work with a wide range of consultants and professional agencies, in order to acquire the necessary skills for development work. In other words, the vital link between community participation and technical and financial development is being made.

Civic organizations throughout South Africa have also been engaged in local-level negotiations with the authorities, primarily over the future of local government, and the need to unite the divided cities and towns. The civic organizations have put forward new models of local and metropolitan government relying on a single tax base, so that future municipal areas will have more equitable access to finances necessary for the provision of land, services and housing.

In conducting struggles and campaigns at a grassroots level, civic organizations have become particularly aware of the need for a healthy environment, with access to parks, gardens, recreational facilities, open spaces and clean air for all. However, these issues are seen as primarily political, economic and organizational, and not relegated to 'environmental awareness' on the part of individuals.

Garden village or ghetto?

In conclusion, it is important to emphasize that the urban environmental crisis in this country is related to an absence of basic resources for decent human life — land, housing and services. Unless and until these are solved, it will be hard not to see the environmental concerns of the garden villagers — or the northern hemisphere environmentalists — as something of a luxury.

LESLEY LAWSON

ABOVE
The squalor of Alexandra township
BELOW
Paved roads and tidy gardens in Cape Town's northern suburbs

JANNIE GAGIANO

•JAPHTA LEKGHETO•

Japhta Lekgetho, founder of the National Environmental Awareness Campaign (NEAC) and African National Congress spokesperson on the environment, sits at his desk in a simply-furnished, colourless office in Dobsonville, Soweto. Through the window behind him one sees the lush lawns of the people's park in which NEAC's resource and community centre is situated. In the sprawling, smoggy, dusty, litter-strewn environment that surrounds it, this oasis is almost a miracle.

It is Japhta Lekgetho's miracle, one for which he has fought long and hard, and one he would dearly love to see repeated a myriad times in South Africa's black ghettoes.

Once a high school teacher of geography, biology and economics ('In the Bantu education system you have to be a jack of all trades'), Lekgetho, like many others, resigned from his teaching post after the student uprising of 1976. 'I left teaching,' he says, 'because of the inequalities, the unpleasant working conditions.'

In that same year, as services ground to a halt in the strife-torn townships and the accumulation of filth in the streets threatened to pose a serious health hazard, Lekgetho saw a need 'to mobilize people to be more aware of the dangers surrounding them.'

He had long been painfully aware of the difference between the environments occupied by blacks and whites. 'The whites' environments were clean and well-structured, ours were disorderly and uninhabitable.' It was time to change some of that.

His initial problem, he says, was to create more independent thinking in the minds of a people long subjugated to government autocracy and state-created living conditions. Prevented from owning their own property, herded into faceless ghettoes, township dwellers had no stake in improving their environment.

'It's difficult removing people from a position of dependence to a position of independence. You need to apply tactics that will work with the mind. You need to discuss certain issues that are crucial to their lives. It was a slow, painful progress but today one is happy that people are more mentally independent,' he says.

Lekgetho's aim was to encourage mass participation in a project which would involve and affect the whole community. It was clear that the place to start was with the huge body of largely disaffected youth out on the streets, their schools closed after the uprising. And so Operation Clean Up was born.

As young people were taught to clean up their own areas, the message of a cleaner environment was spread. The operation, maintains Lekgetho, also taught people about their civic rights, about property ownership, pride and the right to live in a decent environment.

Towards the end of 1978, NEAC was formed as a vehicle to promote environmental awareness. It has, claims Lekgetho, 'worked wonders. The clean-up operation became a revolution.'

But success brought equal amounts of frustration, largely caused by the inability of impoverished local structures to cope with the demands. 'The local councils were unable to provide regular services. So you are aware that you live in a dirty environment but you don't know what to do. While the people play their role the local council can't play its role.'

He blames apartheid, with its inequitably distributed resources of land and wealth, for much of the damage to the environment and, though he does not see the damage as irreparable, he believes time is of the essence and that time is running out.

'While one accepts that there have been a lot of announcements that apartheid has been removed and one would be very happy to see

that being practically done, the process is slow while the destruction of the environment isn't slow.'

There is a need, he believes, for a change in attitude of both blacks and whites. 'Whites must learn to share with other people and to accept the fact that these are the indigenous people, and blacks must accept that whites are also South Africans.'

He sees as non-negotiable, if the environmental future is to be assured, a redistribution of land — 'already the 13 per cent of land the blacks have is overloaded and the resources have been exhausted by deforestation and erosion. In the urban areas people are overcrowded, services are outmatched. We need a well-structured, non-racial, democratic policy on the environment that's going to protect our natural resources and encourage wise management of resources; a policy that will make the resources accessible to all the people; a policy that's going to make development economically and in other facets equitable to the environment.'

Lekgetho believes it imperative that a charter or code of protection of the environment be incorporated in any new constitution. 'My belief is a Bill of Rights isn't formulated in isolation, it should be relative to the environment, very holistic.'

He is also very aware that the protection of tomorrow's environment depends on today's youth. It is with that aim that NEAC's resource and recreation centre, set in this Dobsonville Park, was established. Young people are encouraged to make use of the growing library of books on the subject as well as to enjoy the facilities of soccer fields and picnic places. And out in the community, there is a drive to plant, to green the grey townships.

But Lekgetho continues to hope; continues to hold meetings with members of the community; continues to press for an intensification of clean-up operations and the beautification and development of the parks.

'We are looking at a new South Africa with a new positive attitude. We will intensify our efforts and involve people at grassroots level. We have to work towards building confidence and hope that things will be fine, and we are the only people who can make them fine,' he maintains.

Part of that education drive will be to help people to see the situation holistically. 'People aren't aware that if you speak of the economy you speak of it taking place in an environment, in a structure that's well-planned. People see starvation in isolation, not as part of the misuse of land. If only people could understand that if you protect the environment you can get food, housing, prosperity.'

Lekgetho is convinced that that day will come. Describing himself as a realistic optimist, he speaks of the shift that is taking place and the need to direct that shift without 'competing with the interests of the people. The aim should not be disruption, you must be persevering and understanding and you need financial resources as much as you need human resources.'

There is a lot to hope for and a lot to do and Japhta Lekgetho will keep on doing what he can. His years of working virtually alone have been recognized by his appointment as ANC spokesman and that has given him the strength and the confidence to keep going. Still, he admits, 'I don't have a bag full of solutions, we must work through the problems together'. ❐

5

VICTIMS OR VILLAINS?

Overpopulation and environmental degradation

◆ BARBARA KLUGMAN ◆

'Development is the best pill. Take care of the people and population will take care of itself' (Berer, 1987:5).

An all too common assumption is that overpopulation is a major cause of environmental degradation. Moreover, it is argued that since overpopulation is the cause of the problem, lowering the population growth rate is the logical solution.

In this chapter I argue that all societies need a balance between economic resources and population growth and that in South Africa, it is the process of colonialism and apartheid that has destroyed this balance and thereby caused environmental degradation. In particular, it is the inequitable access to power and use of resources that threatens our environment. It follows that the solution to the environmental threat lies in restoring the balance between population and resources by redressing these inequalities.

The question of population growth has unfortunately been highly politicized in South Africa. At the one extreme we have the victim-blaming position, which argues that the poor are responsible for their own poverty because they have too many children. At the other extreme is the view that women ought not to use contraception because by having more children, a group can claim more power for itself. This position has been developed in defence of both the 'Afrikaner nation' and the 'African nation'.

It is incorrect to see the wealth of the First World, and the poverty of the Third World, as simple facts which bear no relation to each other

Women are at the fulcrum of both of these positions. They are also relatively powerless, both in their domestic relations and in relation to political decision-making. This is doubtless part of the reason why the central thrust of the environmental debate has been around control over women's reproductive abilities — in other words, population control. This skirts around and trivializes the critical question concerning South Africa's environment — the question of how its resources are being and should be deployed.

In particular it must be remembered that many of the major environmental problems facing South Africa today — for example, toxic waste and acid rain — have little or nothing to do with population size. This chapter deals with those environmental problems that are often blamed on population size: overcrowding, soil erosion, deforestation and the like.

The central question, I would argue, is that of the use and misuse of resources. Who decides about the priorities for resource usage? Who actually uses up resources?

Population and the environment in a global context

'When we talk about contraception it's just another way of murdering us. We have been extremely exploited and insulted by past policies. For instance, methods such as depo-provera — which are banned in many parts of the world — have been pushed onto us' (Lydia Kompe in Collinge, 1990:3).

'In Africa — as in Asia and elsewhere — overpopulation spells abject poverty. Until people are persuaded to adopt a more responsible attitude towards parenthood, they will have to live or die with what flows from their folly' (Delph in *The Star*, 7 September 1990).

'The conventional wisdom about the population question is that overpopulation arises because of the scarcity of resources available for meeting the subsistence needs of the mass of the population' (Harvey, 1974:272).

It is a common assumption that it is people in the Third World who put most pressure on the land and are the major cause of the global population problem — this because they live in countries with collapsing economies which are not growing fast enough to keep up with their population growth rates.

The overpopulation view, while empirically correct, makes no

BELOW

Flush against the main road in Khayelitsha, this squatter community grows daily as the influx of people to the western Cape continues to grow. Roads are planned but before building starts, people start setting up housing on the open spaces which are close to water and transport amenities

JANNIE GAGIANO

'Development is the best pill. Take care of the people and population will take care of itself'

attempt to analyse the reasons for matters such as the inadequate food production in these countries — or, more fundamentally, the reasons for the overall economic collapse of countries in the Third World. It leaves us with the conclusion that population density is the problem (see Box A) and that tackling the population growth rate is the only way out of this poverty.

> It is necessary to emphasize ... that almost every facet of environmental quality, and of sociopolitical and economic development, is a function of human numbers. Human numbers could kill the world (Huntley, Siegfried and Sunter, 1989:25).

This view fails to examine the nature of resource usage globally, suggesting too simply that Africa cannot feed itself and, more seriously, that the rate of population growth in the Third World is responsible for the overutilization of international resources, and hence the threat to the environment. Huntley, Siegfried and Sunter express this as follows:

> It is not only the absolute numbers that are cause for alarm, but also the rapidity of growth during the past century and the massive disparity in age and wealth between the so-called Developed and Less-Developed Countries (1989:22).

They say that 76 per cent of the world's population lives in the Third World, which generates only 19 per cent of the world's Gross Domestic Product (GDP) and has a population growth rate of 2,2 per cent per annum. The First World, in contrast, has 16 per cent of the world's population and accounts for 69 per cent of the world's GDP, while it has stable or decreasing populations (1989:22-23).

BOX A

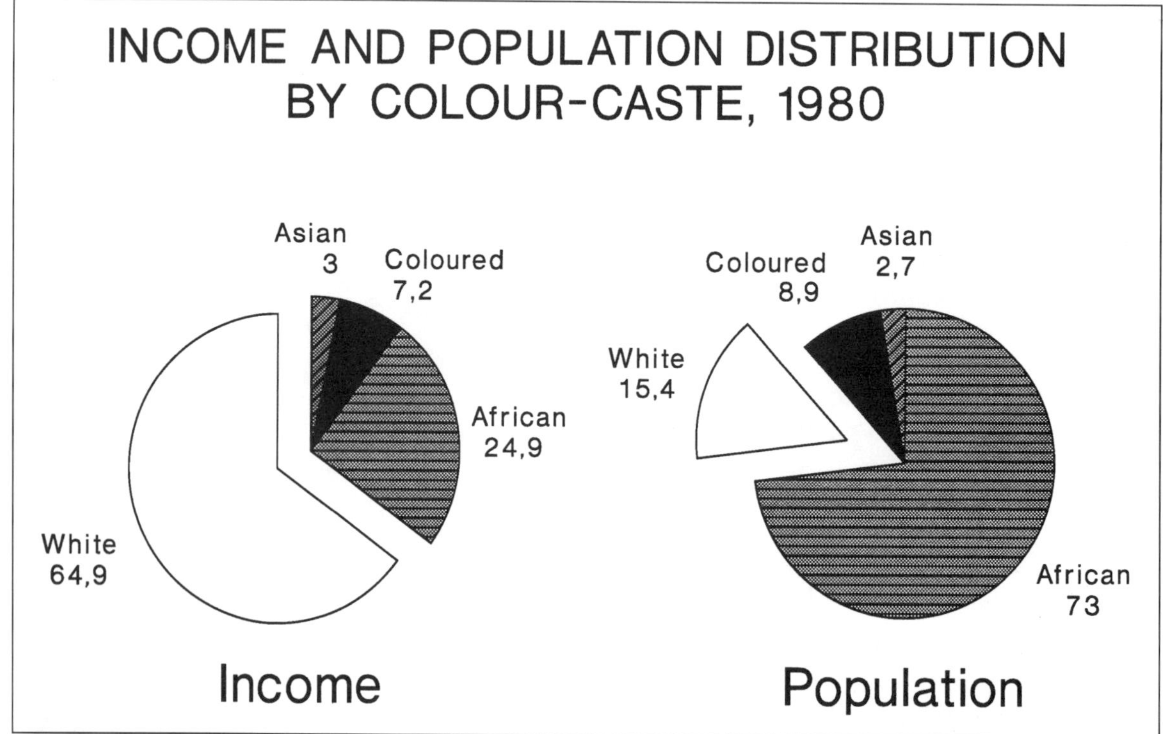

What is left out of this description is the fact that the relative wealth of the First World derives directly from its use of Third World resources and Third World markets on terms of trade which have always been advantageous to the First World. It is incorrect to see the wealth of the First World, and the poverty of the Third World, as simple facts which bear no relation to each other. There are a number of central issues which need to be understood in assessing the question of who is using the world's resources and who is threatening the world's environment. First, there is the question of the terms of trade on which the Third World is incorporated into the international economy; then there is the question of the Third World debt and the capital flows relating to it; and finally, the question of the amounts of resources consumed.

The colonial economies of the Third World were developed in order to service European economic needs and were incorporated into the international economy as providers of raw materials and cash crops. This long-standing international division of labour traps Third World producers in a position of perpetual subordination from which they cannot escape (Campbell, 1989).

Today, debt servicing increasingly uses up the GDP of Third World countries, which should be used for economic development and environmental protection. Long-term and short-term liabilities in sub-Saharan Africa went from 30 per cent of the region's combined GNP to 50 per cent from 1978 to 1984.

Thus the economies of Third World countries are structurally subordinated to those of the First World. Their economic weakness locks them into a position of perpetual underdevelopment. To blame Third World population numbers for poverty and for putting undue pressure on the environment is to blame the victims of this process.

The argument that the poor people of the Third World use proportionately more of the world's resources, while contributing less to the world's GDP, compounds the victim-blaming syndrome. People in the First World consume more resources than those in the Third World, as illustrated in Box B. Redclift argues that

> It is intellectually dishonest to attribute the global resource crisis to the population explosion in the less developed countries, without acknowledging that the share of resources consumed by poor people in these countries is much smaller per capita than it is in a country like Britain (1984:23).

Overall, the populations of the First World consume two-thirds of the world's food production (George and Paige, 1982).

Third World countries have to export their food products in order to gain desperately-needed foreign currency. For example, pigs and poultry in Europe are fed 6 million tons of fishmeal each year. Fishmeal for animal feed has become a major export from Peru. As a result, fish in Peru has become too expensive for the people, even though it is an essential source of protein in their diets (Dumont and Cohen, 1980). This also applies in South Africa, where there is a high rate of malnutrition but food is exported (Scholes, 1989; Science

It is the inequitable access to power and resources that threatens our environment

Committee of the President's Council, 1983).

Deforestation is occurring in Brazil in order to clear land for cattle ranching; the cattle are produced for meat which is exported to the United States of America. It is Brazil's need for foreign currency which is leading to the process of deforestation. Many of the global environmental problems we face today arise directly out of the position of Third World countries in the international economy.

Harvey (1974) notes that many of the pressures on the world's resources and environment could be reduced if rich people, largely in the First World, changed their consumption habits. The centrality of meat in the American diet is a case in point. As the examples above indicate, the production of meat for the United States of America has led directly to deforestation in Latin America; the production of fishmeal and grain to feed livestock uses up food which could otherwise be available to malnourished people. Adequately nourishing a world population that in 2030 will be 60 per cent larger than today's, will preclude feeding a third of the global grain harvest to livestock and poultry, as is currently the case.

Unless the problem of the subordination of Third World economies is resolved, and unless the excessive consumption of resources in the First World is decreased, further environmental degradation cannot be prevented.

Population and the environment in South Africa

'The human population explosion ... is the most critical environmental problem in southern Africa' (*Quagga*, September 1990).

Resource usage within South Africa

In South Africa the poor are held responsible for the consumption of resources, and hence for putting pressure on the environment through activities such as destroying trees for firewood, overgrazing and producing waste. In addition, since the African birth rate is higher than the white birth rate, it is argued that the future depends on lowering the growth rate of the African population:

> South Africa finds itself in a unique position, demographically speaking. The various population groups are at different stages of demographic transition (ie phases of population growth). The developing population groups of South Africa are well into the pre-modern phase, displaying typical features of it such as high population growth, poverty and illiteracy. The more developed section of the South African population has already moved through this phase into the more modern phase.
>
> The South African population is currently growing at an average rate of 2,3 per cent per annum (whites 1,76 per cent, coloureds 1,8 per cent and blacks 2,8 per cent). If this trend were to continue, the indispensable balance between population size, socioeconomic capabilities and available resources of South Africa would be upset, with far-reaching social economic consequences that could seriously jeopardize stability and progress. (Department of National Health and Population Development, 1987:1-2).

'Contraception is just another way of murdering us. Methods which are banned in many parts of the world have been pushed onto us'

— Lydia Kompe

BOX B

Population density

The table on population density shows us that countries which we think of as overpopulated, such as India and China, are no more populated than countries in Europe. People living in countries with a high population density are not necessarily poor and do not necessarily undermine their environment. Numbers per se are not the problem: the problem is access to resources.

Country	People per sq. mile	Continent	People per sq. kilometre
France	251	Europe	101
China	271	Asia	102
Holland	1117	South America	15
India	516	Africa	18.2
Great Britain	583	Africa excluding arid land	39

(Harrison 1897:242)

This concern for South Africa's future population must be seen in relation to the poverty and suffering of the majority. The overpopulation argument fails to consider the history of the distribution of resources in South Africa to a tiny minority, and the denial of resources to the majority — whether in the form of access to land or to education or to social and political opportunities. It is this denial that has already upset the balance between population and resources, and hence between population and the environment. Not only has it created massive inequalities in the use of resources, but it has also resulted in a high population growth rate.

> It's the laws which make us overcrowded, laws which limit blacks to 13 percent of the land. We haven't got the land, so whether we stop giving birth or not we will be overcrowded (Lydia Kompe in Collinge, 1990:3).

The creation of an overpopulation problem in South Africa

> It has been estimated that in the twenty-three years from 1960 to 1983 a total of 3,5 million people, almost all of them black, have been subject, in terms of government policy, to forced removal from one place to another where they did not choose to go (Wilson and Ramphele, 1989:216).

This has occurred in five different ways: through Group Areas Act removals; through the expulsion of people from urban areas; through 'black spot' or Bantustan consolidation removals; through 'betterment' programmes to consolidate populations living scattered through the reserves into villages; and through forced removal off white-owned farms (Wilson and Ramphele, 1989).

'Until people are persuaded to adopt a more responsible attitude towards parenthood, they will have to live or die with what flows from their folly'

> At the beginning of the 1950s one-third of black South Africans lived on the white-owned farms, which constitute approximately 80 per cent of the area of the country ... By the end of the 1960s the absolute number of jobs in agriculture started to fall. More and more people were pushed off the land. There is nothing unusual about the displacement of agricultural workers through mechanization; many other countries have experienced it. What was unique about South Africa was the response of the state to these changes. Africans evicted from the farms were not allowed to move to the cities. They had to go to the homelands, to the already overcrowded labour reserves. It is estimated that between 1950 and 1980 some 1,4 million people were squeezed off the commercial farms and another 90 000 were pushed off or migrated from towns other than the eleven metropolitan areas in South Africa. Of these 1,4 million ended up in the reserves. (Simkins, 1983:59-61 cited in Wilson and Ramphele, 1989:223).

QwaQwa: a case study

> In QwaQwa (an extreme, but by no means unique, example) the population in 1970 was estimated at 24 000, almost five times what it had been during the First World War when a government commission had found it already overcrowded. But by the mid-1980s, through a combination of agricultural mechanization and the application of an ideological policy to reduce the number of blacks on white-owned farms, the population had risen to approximately 500 000 (Wilson and Ramphele, 1989:223).

Niehaus describes how in the 1960s Radio Sesotho attempted to lure people to QwaQwa:

> The radio said: "Come and enjoy yourself in QwaQwa. All the black people must come to this homeland where they will find fields and lots of jobs. Your chiefs will look after you" (1989:164).

However, reflecting on his experience on arriving in QwaQwa, one of Niehaus's informants said:

> The radio lied. Here there was no grazing for cattle and people had to wait a long time for work. (ibid:166)

Referring to the closer settlement of Tseki, in QwaQwa, Niehaus describes the increasing pressure on water and fuel resources:

> There was no clean water in Tseki in 1974 and, in the absence of taps, people were forced to fetch water from a few wells on the nearby mountain slopes. Informants recalled that people got rashes, and sores in their mouths, from drinking the well water, which rapidly became polluted ...
>
> Taps were provided late in 1975. Although the taps were few in number, and fetching water was still an arduous occupation in the 1980s, people noted the immediate improvement in health and hygiene which resulted from the provision of clean water.
>
> People who had been used to a free and abundant supply of firewood on the farms were appalled by the lack of fuel. Trees in the area were rapidly consumed, and people had to go into the mountains to fetch wood. After a few years small businesses sell-

JUSTIN SHOLK

ing wood and coal were established in the area, but in 1983 a small bundle of wood ("enough for two days") cost R2,50, and the price of a bag of coal ("which lasts a week when it is not too cold") was R4,00. Most households could not afford to spend R12,15 a week on fuel, and women were still walking to the mountains to bring back wood. By the 1980s people had to go all the way to, and sometimes over, the Lesotho border to find a ready source. Women banded together to make a journey, and armed themselves with sticks and axes, claiming that "dissidents" from Lesotho had attacked and raped women collecting wood on their own in the past. The round trip took more than eight hours, and on numerous occasions women were stopped by the headman or his induna as they re-entered the closer settlement, deprived of their wood, and fined for "creating tensions on the border" (ibid:167).

ABOVE
A woman in Transkei carries a pile of the only fuel that is locally available

This case study of QwaQwa illustrates a number of general trends which help us to understand the reasons for both the overcrowding of the 'homelands' and the concomitant environmental destruction.

First, the high population densities of the 'homelands' reflect the apartheid design. There is overcrowding because people have been forced to move into the 'homelands' instead of being allowed to remain on the land on which they were born or move to urban areas. The apartheid government must take responsibility for overpopulation in these areas, not individuals.

Second, the 'homelands' lacked a basic infrastructure in the form of adequate sanitation and sufficient water and fuel, such as electricity. As a result there has been rampant destruction of available resources in the form of soil erosion, water shortage and pollution, not to mention deforestation. Again, the solution lies in a major development effort in rural areas.

Given that all societies require a balance between population and resources, we have to account for the high population growth rate. I shall argue that apartheid has not only led to distorted population densities, overpopulation and environmental destruction in certain areas, but also to a high population growth rate among those people who have been hardest hit by apartheid policies.

Causes of the high population growth rate

The overpopulation theorists argue that high population growth rates are a result of the process of demographic transition — the population growth rate increases as a population shifts from 'natural fertility' to 'high fertility rates and relatively low mortality rates because of access to First World medicines and infrastructure...' (Huntley, Siegfried and Sunter, 1989:49). There are two inaccuracies contained in this view, both of which obscure the real causes of population growth.

The destruction of social controls over reproduction

The first mistake is the idea that there has ever been 'natural fertility'. People have always controlled the process of reproduction. McLaren points out that

> Anthropologists have uncovered, in pre-industrial societies, a vast range of rules, regulations, taboos, charms and herbal remedies for the purpose of affecting the processes of conception and gestation (1984:2).

Prior to industrialization, this held true for the African population in South Africa. 'Traditionally', African population growth was regulated by a range of cultural rules including the practice of intercrural sex (sex without penetration) and prohibitions on full sexual relations before marriage, for a prolonged period after the birth of a child, and sometimes for the entire breastfeeding period.

The balance between population size and resource usage was undermined by the form that industrialization took in South Africa. Migrant labour in particular dramatically affected the African population's control over reproduction, because of the separation of husbands from their wives and the undermining of normal social relations. The disruption of family life and the breakup of viable and stable social relations led, among other things, to substantial changes in sexual mores. For example, 'traditionally', adolescents were taught accepted sexual morality and behaviour at initiation schools. They were allowed to be sexually active, but could engage only in intercrural sex. Their compliance was monitored by older members of the society. This social institution has largely collapsed, and has not been replaced by other forms of sex education and social control over sexual behaviour.

Similarly, contact with Europeans, and particularly with Christianity, shortened the duration of breastfeeding, which in turn shortened the period of intercrural sex or abstention (McKenzie, 1988). Thus another cultural practice which limited population growth was undermined over this period.

The truth about mortality rates

The other misconception held by the overpopulation theorists is that the population growth rate among Africans is high because the mortality rate has dropped through access to modern medicine.

ABOVE
Children at Sada in the Queenstown district, Ciskei

However, McKenzie convincingly demonstrates the inaccuracies and oversimplifications contained in this assertion. On the question of illness, for example, he shows that

> as poverty deepened in the society not only did the diseases of poverty commonly experienced in Africa (eg malaria) increase, but a whole new burden of diseases was added (eg smallpox, tuberculosis). Many of these diseases (eg tuberculosis) are intimately related to poverty (McKenzie, 1988:109).

Moreover, access to health care is racially determined: there is approximately one doctor for every 12 000 Africans as compared with one doctor for every 330 whites, and only 5,5 per cent of doctors practice in the rural areas. In addition, the creation of poverty in South Africa has led to a situation in which the death rate for black children from nutritional diseases is 31 times higher than that for white children

Giving women access to education is critical to their empowerment and is the major factor leading to a decline in fertility

(Omond, 1985:74). Keenan and Sarakinsky (1987:108) argue that even in the last decade the standard of living of the majority of the African population has declined.

> The food we eat is not the same. We have given up drinking milk. You find a young child telling you she is tired which did not happen with us before (Informant cited in Klugman, 1988:89).

Both the birth rate and the death rate have been directly affected by apartheid. Ironically, as discussed below, it is poverty and lack of security that lead people to have many children.

'Children build the nation'

'Children build the nation'. This statement (informant cited in Klugman, 1988:87) highlights the fact that in a situation of poverty, having many children is often the most rational choice. People need children.

Children play an important role in rural life. Given the absence of men and the heavy domestic burden carried by women, for example in the hours spent in fetching water, children carry out many of the daily tasks. Employed women often rely on their daughters to help them with the housework (Cock, Emdon and Klugman, 1983).

In an extended family situation, children usually contribute to the family income. Thus in the present situation of widespread unemployment, the more children one has, the greater the chance that at least some will gain employment.

Both as a cultural pattern and because of the lack of adequate social security benefits, poor people rely on children to support and look after them in their old age. 'Children are of help when not well; when old they keep you company,' and 'A person must have somebody to care for you when old' (Informants cited by Klugman, 1988:87).

Having only one or two children is no guarantee that a rural African person will see any children live into their adulthood, as a result of the high infant mortality rate. And until such time as those factors which cause high infant mortality change, people will continue to have many children (Sai, 1986). 'To date no country has achieved a low birthrate as long as it has had a high infant mortality rate' (Hartmann, 1987:9).

'Women were created for childbearing'

Finally, for most African women, motherhood is their primary source of social validation and self-esteem, as the above statement suggests (informant cited in Klugman, 1988:87). It is necessary for women to have children in order to receive respect, status, and full adult legitimacy both in their own eyes and in the eyes of their peers.

Most women have no other options. They have never had the educational and social opportunities which would allow them to develop careers from which they would derive satisfaction and social status. Internationally women's lack of education and lack of control over their lives is considered the central factor causing their high fertility. They will not have fewer children until they have more control over and more choices in the rest of their lives.

Conclusion

All societies need a balance between resources and population. In South Africa, that balance has been destroyed by apartheid policies and through this process, major environmental problems have been created.

It is easy for these problems to be blamed on overpopulation, but this approach is both ahistorical and dishonest. It is not population numbers that threaten South Africa, but the lack of access to resources on the one hand and the overconsumption of resources on the other. It is not the poor themselves who have caused their poverty by having many children, but the practice of discrimination — which has denied them access to resources and security and in so doing, has caused the birth rate to increase.

Nowhere in the world has the mass provision of contraceptives led to a decrease in the population growth rate in the absence of an improvement in women's position in society, and an improvement in the overall standard of living. It is now internationally recognized that as people's standard of living improves, and as they are able to exercise increasing control over their lives, so they choose to have fewer children (United Nations, 1975.)

Two specific and crucial factors in improving the overall standard of living of the population are education and health care. In particular, giving women access to education is critical to their empowerment and is the major factor leading to a decline in fertility (Sai, 1986).

Nowhere in the world has the mass provision of contraceptives led to a decrease in the population growth rate in the absence of an improvement in women's position in society

Likewise, a central part of primary health care services is the focus on mothers and children — the provision of health education, contraception, infertility services, immunization, and ready access to clean water. All of these enhance the quality of people's lives and as a result, lead to a decline in fertility.

This understanding of the population-resources question leads me to conclude that what South Africa needs is not a population policy, but a resources policy. The solution lies, among other things, in a concerted and democratic effort to use South Africa's resources in a constructive, environmentally conscious manner. Such a national development policy would improve not only the standard of living of the poor, but also the degree of control which they have over their lives.

BELOW
A malnourished child in a rural clinic

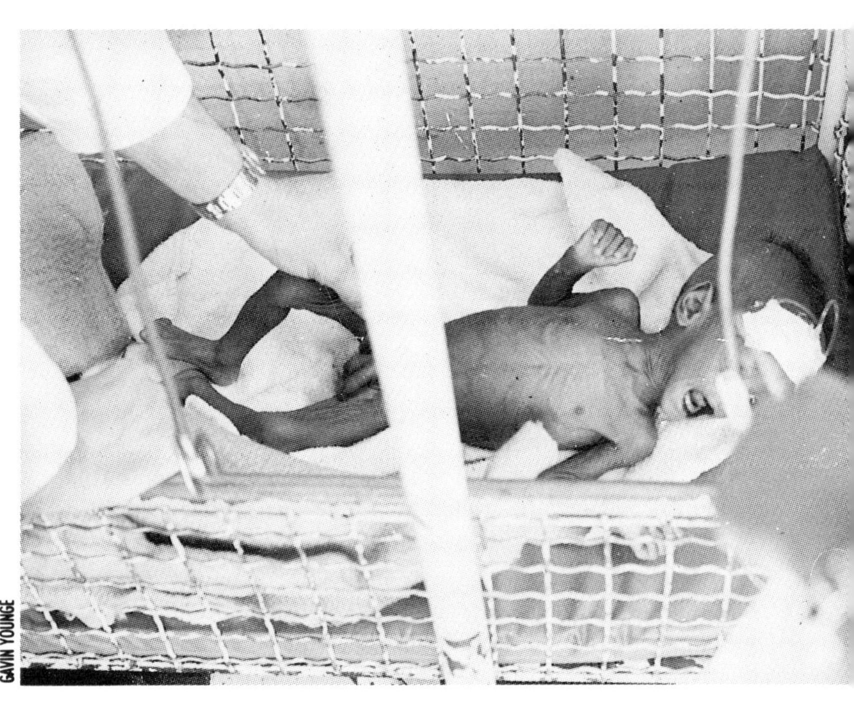

GAVIN YOUNGE

6

REDS AND GREENS

Labour and the environment

◆ ROD CROMPTON ◆ ALEC ERWIN ◆

Reds and greens

The struggle for a fundamental restructuring of the social and economic dispensation in South Africa has a long history. This red cause has drawn its inspiration from a large body of socialist theory and history. It is a cause that now has to reflect on the experiences of the USSR and Eastern Europe and integrate these lessons into a new approach to South Africa's future.

The struggle to protect the environment from the ravages of humankind is more recent in South Africa and it draws on the green movement that has also more recently developed in Europe and the United States. In South Africa there has of course been a concern for wildlife and soil conservation, but the growth of a wider environmental concern undoubtedly owes its origins to the success of the green movement in other countries.

In Europe red and green politics have cooperated on certain issues but largely remain separate. Increasingly, however, this divide seems problematic. The reds are becoming aware of the importance of the environment — the effects of Soviet central planning on the environment providing a stark lesson — and the greens are becoming aware of the role industrial planning plays — Chernobyl and Bhopal being harsh reminders here. A recent work defines the link very clearly:

> Economic activity is precisely about using resources (taken from the environment) to produce goods and services, with resulting wastes (deposited back into the environment). If environmental considerations are not fully integrated into economic policy, there is no hope of tackling the ecological crisis (Jacobs, 1989:11).

We want to argue that in South Africa this integration of red and green political traditions is essential. Repressive regimes such as that of apartheid would seem to be the greatest enemy of the environment because they crush all opposition to the effects of their decision-making. In rebuilding our society we not only have to

repair past damage but also have to ensure that we minimize future damage to the environment.

The thrust of our argument is that both the red and green traditions of politics require a vibrant democratic society where the state leads economic development and where strong organs of civil society protect and advance legitimate social and economic interests. A red-green alliance is essential for our future.

The starting point for arguing in favour of this alliance is an examination of a question whose answer is not as obvious as it may seem. This is: where is the environment? This question is particularly pertinent for South Africa.

The 'work environment' is seen as somehow separate from 'the environment' — but for many people, industry *is* their environment

Where is the environment?

'The environment', as it is often conceived of in our media and particularly in 'coffee table' publications, is the sky, the oceans,

GAVIN YOUNGE

LEFT
Workers in a steel foundry — without protective clothing

wildlife and game reserves. It is what industry does to this environment that is held to be of concern. This conception ignores the fact that for many people, industry is their environment. Somehow a distance, a separation creeps in between the here and now that ordinary people experience, and 'the environment'.

This notion of separation is an extraordinary one. Much of the damage to the environment begins at the point where one form of energy is converted to another — for example, where coal is converted to electricity, or oil to petrol. At almost all of these 'energy conversion points' there are people attending to the process. For example, where SASOL converts coal into petrol thousands of workers are employed to make that process happen: yet the employers who design these processes and the workers who carry out their instructions are not seen as acting or working in 'the environment'.

Instead, workers are said to work in the 'work environment' which is seen as somehow a separate environment from 'the environment'. If this false separation between the 'work environment' and 'the environment' is removed then a whole new way of looking at the environment becomes possible. In this revised view we recognize that environmental damage begins right at the point of production — and thus workers suddenly find themselves on centre stage right under the spotlight. Industrial democracy immediately becomes a vitally important issue in transforming dirty industry into clean industry.

Almost all toxic substances make their presence felt first of all in the workplace. Traditionally most South African employers have responded to industrial hygiene problems not by eliminating the source of the problem, but by insisting instead that workers wear protective clothing. If workers must wear personal protective equipment then by definition environmental contamination is taking place.

The safety and environmental risks arising from bad plant design, poor plant maintenance, dangerous operating procedures, the handling of unlabelled hazardous substances or the use of untrained contract workers will threaten the individual workers in the workplace first of all. The first communities to feel the effects will be those of the working poor who are usually located closest to the industrial areas.

If workers must wear personal protective equipment, then by definition environmental contamination is taking place

Workers in the environment

If one accepts that all citizens have a right to a healthy and safe environment then it follows that workers have a right to jobs which are healthy and safe. Trade union struggles for health and safety in the workplace constitute the first line of defence for an embattled environment.

Why have the trade unions made so little impact on occupational health and safety and the environment when they have shown themselves capable of strong action in recent years?

In the early 1980s the emerging progressive unions were beginning to take up health and safety issues in a serious way. This movement was nipped in the bud by the introduction of the Machinery and Occupational Safety Act (MOSA) in 1983. MOSA set up safety committees 'which from the workers' point of view are completely toothless' (Meyers and Steinberg, 1983:86). Although some improvements were later made to the Act, it has been largely successful in sidelining trade unions from health and safety issues.

Worldwide routine industrial production has a kill rate that is the equivalent of one Bhopal per day

Industrial democracy: three basic rights

The democratization of the workplace and the removal of obstacles which, like MOSA, prevent unions from taking up safety and environmental issues, offer one of the cheapest insurance policies for the environment whatever the economic growth path chosen by society.

However, the key to successful union action to protect the environment from the detrimental impact of industry is industrial democracy. This can be distilled to three basic rights:

- The right to know
- The right to act
- The right to refuse unsafe work

ABOVE

The protest against Thor Chemicals united several different groupings

The right to know

Every year approximately 1 000 new chemical substances are added to the current list of between five and six million. The public who buy them and most of the workers who produce or transport these toxic substances are kept in the dark about their real nature.

Details of plant design are equally crucial. As industrial installations get bigger and bigger the consequences of catastrophic plant

Government legislation has made it a dismissable offence to refuse to do something which you think will kill you

failure or accident become more and more serious for the workforce and the surrounding community.

The worst industrial accident known took place at Bhopal in India in 1984 (Kharbanda and Stallworthy, 1988:17). On the night of 2 December a leak of deadly methyl isocyanate gas from Union Carbide's pesticide factory killed 3 677 citizens. A further 80 000 were severely affected and 341 000 eventually registered compensation claims for disablement, injury, loss of property, and loss of earnings resulting from the disaster (Herding, 1990). If the community had been informed of what might happen in the case of a disaster, the simple expedient of wet cloths may have saved many lives.

According to a director of the United Steelworkers of America, worldwide routine industrial production has a kill rate that is the equivalent of one Bhopal a day — approximately one million people per annum (personal communication, 1990). We live and work in economic systems which are extremely hazardous to the health of both the environment and the people who live in it.

An explosion at SASOL's giant Secunda complex in 1989 killed fourteen people. In the official inquiry into this accident SASOL refused to disclose the plant design to the union. Will there be another, bigger explosion? Should workers and communities have to put their lives in management's hands without knowing the risks involved?

The right to act

The right to act follows naturally on the right to know. This includes the right for concerned groups like unions and green organizations to inspect and report on polluting or hazardous industrial installations and to negotiate controls where necessary.

The right to refuse unsafe work

The right to refuse unsafe work is a vital issue for workers. Government legislation has made it a dismissable offence to refuse to do something which you think will kill you. 'Unsafe work' in our definition includes work which is unsafe for the environment. The Finnish paper workers' union won the right for a worker to refuse to dump anything which he or she considers to be hazardous waste (ICEF, 1990).

Workers and the community

The development of active democratic organs of civil society can play an important role in defending the rights of workers, the community and society as a whole from the onslaught of unchecked economic development. The case of Thor Chemicals, which is discussed elsewhere in this book, most notably in Chapter 11, points towards what can be achieved.

Thor Chemicals hit the headlines, locally and internationally, in April 1990 when massive concentrations of highly toxic mercury were detected in the Umgcweni River near its Cato Ridge plant in

Natal (*The Daily News*, 5 April 1990; *The Sunday Tribune*, 8 April 1990; *The Weekly Mail*, 5 to 11 April 1990; *The Guardian*, 12 April 1990). Mercury can destroy the central nervous system, cause birth defects and a variety of other problems (International Labour Organization, 1985:1332-1338). An analysis of river sediment samples showed that in one case the river sediment was 8 810 times over the limit necessary to classify it as 'hazardous' in the United States (Greenpeace, 1990:5). There were also reports of two Thor workers at the plant having 'gone mad' (*The Guardian*, 12 April 1990).

Thor Chemicals, a British-based multinational, imports mercury waste to South Africa. One of the suppliers of this waste was a company called American Cyanamid. Why did Thor, a British company, decide to build the world's largest toxic mercury recycling plant on the borders of KwaZulu in a fairly remote part of South Africa? Why not build it closer to the sources of the waste mercury in the United States or in Europe? We cannot be sure. We do know that such a plant would be illegal in the United States. Presumably the international toxic waste trade is aware of the low wages and very poor environmental protection regulations in South Africa. The international trend is for the industrialized countries to peddle their waste in Africa, the Caribbean and Latin America (Greenpeace, 1989).

Four groups took up the issue with Thor publicly: Earthlife Africa, the Chemical Workers Industrial Union (CWIU), the local residents under their chief and white commercial farmers from the Tala Valley. (The perhaps unexpected involvement of the latter is probably due to the fact that their vegetable crops had been severely affected by pesticide-spraying by the sugar industry. They were engaged in legal battle against several large pesticide manufacturers as a result of this.)

The CWIU and Earthlife mobilized their international contacts and in April 1990 joint protest demonstrations took place at Thor and at Cyanamid's Bound Brook plant in the United States. (Ironically, while a peaceful demonstration took place outside the Thor plant under the watchful eyes of the security police, demonstrators in America were baton-charged and some arrested.)

These demonstrations were 'firsts' in a number of respects. An embryonic alliance emerged of trade unions, rural peasants and green groupings. The action also had a strong international component: workers and concerned citizens in two countries thousands of miles apart had combined forces to send a clear message to multinationals and the international toxic waste trade: 'You will not go unchallenged!'

The Department of Water Affairs ordered Thor Chemicals to suspend operations for three to four weeks 'due to the recent heavy rains' (*The Daily News*, 10 April.) The Chairperson of Earthlife Africa believes that this is the first time in recent history that the state has forced an employer to close a business for environmental reasons (Albertyn, personal communication, November 1990). Subsequently Thor was allowed to continue operating its plant and was given a new licence to import toxic mercury waste.

In what may be a related development, it was announced that

The trade unions face a choice: they can either adopt a defensive stance as market forces dictate industrial restructuring, or they can take an offensive position and put forward their own restructuring proposals

South Africa would no longer import toxic waste. However, in a separate announcement it was stated that South Africa had ratified the Basel Convention on the storage and transport of toxic waste and would become a signatory to the Convention once local legislation had been adjusted. Why should it have been necessary to sign the Basel Convention if toxic imports were to cease? Is it perhaps because in future they will be imported under the guise of 'raw materials'?

We are still in the early stages of the active defence of our environmental rights by unions in alliance with other groupings. However, the trade unions have in the past demonstrated an extraordinary ability to learn very quickly how to utilize the few rights at their disposal. The emerging green movement in South Africa is demonstrating a great deal of enthusiasm, and this augurs well for the future.

This potential for joint action within civil society to protect the environment needs to be placed in a wider context — a context that reinforces the link between the workplace and the environment. Among the many issues to be examined here are two related ones, namely, the growth path that the economy will follow in future and the link between jobs and environmental protection.

The economic growth path

Apartheid policies have had an exceedingly damaging effect on South Africa's environment. The heavy emphasis on industrialization, with little regard to its environmental impact; the enforced concentration of black people in areas with appalling living conditions, both rural and urban; the lax attitude to toxic substances — all have contributed to environmental degradation (Gelb, 1991).

Paradoxically, this degradation may have been eased to some extent in recent years by the very low growth rate of the economy. However, it is also possible that in some quarters, the reaction to this economic crisis has taken the form of a growth-at-all-costs approach. Environmentally unsound projects abound in South Africa.

However, this low growth rate is creating critical socio-economic problems. There is general agreement that we have to revive economic growth in South Africa, but there is much debate on how this can best be done. There is a free market lobby that advocates unfettered market forces as a panacea for all problems, but this grouping is very quiet on issues concerning the environment.

However, a more fruitful debate is the one taking place within a broad spectrum of opinion that accepts a mixed economy. Here the areas of difference would be the extent and role of the public sector, and which areas of the economy should be used to promote growth.

The latter debate is increasingly concentrating on more concrete, detailed issues, but as yet no comprehensive economic programme has been spelt out by any of the major protagonists. However, the line of thought in the Congress of South African Trade Unions

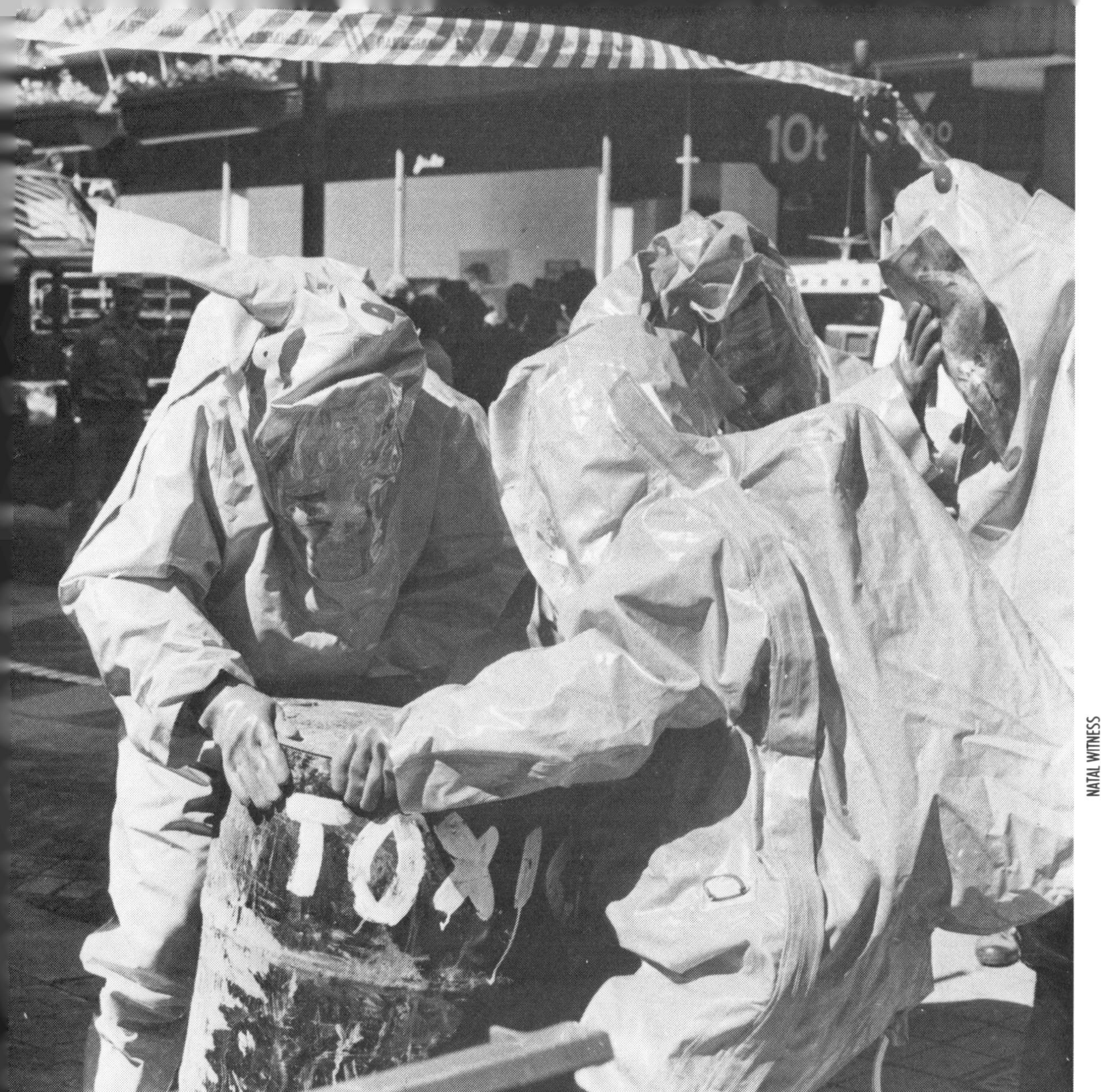

NATAL WITNESS

ABOVE
Firemen in protective suits with their own air supply move drums of toxic waste that were mysteriously dumped in the centre of Pietermaritzburg one morning in 1990

(COSATU) is relatively clear, and at this stage the latter grouping appears to be very close to the ANC on these issues. There has, in fact, been a high degree of interaction in policy formation.

The prospects for economic growth within the limits of the present structure of the economy are not good. Most analysts agree that with favourable circumstances, a 3 per cent per annum growth rate is the most realistic possibility. This rate is clearly inadequate to the pressing social and economic needs of the country.

The thrust of COSATU's thinking is that the state — through decisive but carefully selected interventions — will have to lead the economy onto a new growth path. This requires a coordinated and coherent approach to economic policy. For COSATU there are a number of basic aims that should constitute the core of this programme.

These are: to address the basic problems of poverty, unemployment and shortages of basic social infrastructure such as housing; to increase employment, wages and incomes from small-scale business enterprises; to reintegrate our manufacturing sector into world markets on a competitive basis; and to upgrade the skills and capacities of our human resources.

This path has been defined as a growth through redistribution path since it begins to redress the present maldistribution in income by generating employment and income for the poor. This does not preclude redistribution by fiscal measures, but it acknowledges that such a path is not in itself going to solve the problems of the poor: '... the essential issue is not the redistribution of consumption, but the redistribution of investment' (Gelb, 1990:35).

Such a growth path will require the restructuring of many industries so that they increase employment, raise the general level of their production and lower their costs of production in order to meet the needs of lower-income markets and become competitive on world markets. A process of industrial restructuring cannot be painless and thus certain activities will be phased out while others grow. For these reasons, the unions face a choice: they can either adopt a defensive stance as market forces impose these changes or they can take an offensive position and put forward their own restructuring proposals.

The major COSATU affiliates have proposed the latter path and are involved in developing these proposals. What does this have to do with the environment? Do we once again face the danger that the reds will march on, paying little attention to the greens?

Such a divide would be dangerous for these very important reasons. First, if environmental issues are ignored in the formulation of the growth path, the future costs of environmental degradation will strangle the growth prospect. Second, environmental considerations will be forcibly brought to the attention of South African industry by development itself, by international agencies and by consumer resistance, and this will have an impact on employment. Third, it is in the interests of a strong civil society that worker organization and a green movement act in concert and not in opposition to each other.

Trade union struggles for health and safety in the workplace constitute the first line of defence for an embattled environment

Repairing the damage

It is not necessary to say much more here about how the need to repair the environment will impact on the future growth of the economy: South Africa itself already provides a harsh lesson in this. Because the urban and rural decay that apartheid policies have generated have been ignored for so long, we now have to divert costly resources into simply making certain areas habitable, instead of developing them into viable consumer markets that could revive growth. Given the existing damage to the environment, any new growth path that ignored it would rapidly be strangled by the costs of environmental band-aid, let alone restoration. We will pay more attention to the second and third reasons set out above.

Sustainable development

It is now clear that development and growth cannot ignore environmental factors. Economic activity has to be constrained so that it does not go beyond the limits the biosphere can sustain. For an economy to be sustainable, it must be able to guarantee our children at least the same level of environmental benefits that we have enjoyed.

The environment provides three kinds of benefits that are in fact basic conditions for sustaining economic activity. These are:

- *Resources:* These take the form of raw materials and energy.
- *Waste disposal facilities:* The environment assimilates waste products, but only up to a certain point. Once the limit is overstepped, we have pollution.
- *Environmental services:* These include climate and ecosystem stability and go far beyond the common perception of 'environmental benefits' as taking the form of game reserves and so on. For example,

> How would you like the job of pollinating all trillion apple blossoms in New York State some sunny afternoon in late May? It's conceivable, maybe, that you could invent a machine to do it, but inconceivable that the machine could work as elegantly and cheaply as the honey bee, much less make honey on the side (Meadows, 1990).

The world is without doubt becoming a small place. This makes it inevitable that as the green movement grows in strength elsewhere, it will try to ensure that international agencies enforce responsible environmental practices in all countries. South Africa's future growth path will not escape this pressure.

> In short, human health and welfare ultimately rely upon the life support systems and natural resources provided by healthy ecosystems. Moreover, human beings are part of an interconnected and interdependent global ecosystem, and past experience has shown that change in one part of the system often affects other parts in unexpected ways. (United States Environmental Protection Agency, 1990:9).

The possibility of post-apartheid economic reconstruction offers a unique opportunity — not only to begin correcting the inequity of the past by choosing a new economic growth path, but also to choose a new growth path that will lead South Africa away from the 'dirty and dangerous' economy of the past and towards a 'clean and safe' economy for the future. This inevitably raises the question of the link between employment and environmental protection.

Environmental protection and employment creation

If we look a little more closely at what workers are doing, we find that they are dumping toxic waste, polluting rivers or facilitating the emission of tons of pollutants into the air. Workers actually do these things with their own hands — they are not done by the employers. Why do the workers do this? Why don't they object or refuse? The simple answer is that they need to keep their jobs. The laws of apartheid are stacked heavily in favour of employers: as discussed earlier, to refuse to carry out an employer's 'legitimate instruction' can lead to instant dismissal.

Will an economy limited in the extent to which it damages our life-support systems imply even more unemployment than at present? Business usually puts forward the following simplistic formula:

pollution control = job losses
therefore
jobs = pollution

Thus it is maintained that limiting pollution necessitates reducing the number of jobs; but what kind of employment will be possible in a world of poisoned water, foul air, depleted resources and rising temperatures? What is clear is that trying to hold onto jobs in dirty and 'dangerous factories' until public outcry finally closes them down, at which point it will be far too late to make alternative arrangements, is the worst possible option.

It would appear that part of the solution lies in the elimination of many industries which are just too toxic to continue operating. There is a potential for real tension between the workers who incorrectly see their only choice as being between 'toxic' jobs or unemployment, and the neighbouring communities concerned about contamination from the factories. However, such tensions are not inevitable: a great deal depends upon how the trade unions respond to this challenge.

Taking a longer-term view, a more appropriate formula might be:

pollution = no jobs
therefore
pollution control = jobs

In the longer run, there is likely to be greater job security in the rapidly-developing pollution control and clean-up industries, as a new, environmentally sustainable growth path is followed.

Any economic growth path which envisages a transformation to an economy compatible with the environment will necessarily have to be accompanied by a programme of retraining for workers who have developed skills in obsolete industries. There will also need to be an economic safety net to support workers during the period of transition.

How many jobs will be lost and how many will emerge in the new industries as mass transport replaces cars, cities change shape and social reorganization takes place? This question will require much research. South Africa has the advantage of being able to draw upon the experiences of more developed economies which are far more advanced than we are in this area.

The growth path we seek out must be one where pollution control creates employment instead of destroying it

A planned transition from the 'dirty and dangerous' to the 'clean and safe' is the only viable option. It is the nature and direction of investment which will regulate the speed and nature of the transition. Overall limits have to be set within which economic development can take place, and it is around these issues that a red-green alliance could be forged.

Does clean-up legislation simply mean costs for industry and unemployment for workers? In the United States the introduction of the 1990 Clean Air Act is envisaged to cost $150 billion (R375 billion). In addition, annual costs of ensuring compliance will be approximately $21,5 billion (R54 billion) (ICEF, October 1990). However, this money is not going up in smoke. It creates investment opportunities in pollution control equipment and services, and this means more jobs. One industry's blackened pollution cloud is another's silver lining.

For example, SASOL has reduced air pollution by investing in scrubbers to remove hydrogen sulphide from gas streams used to produce sulphur used in fertilizers, chemicals and for export (*Financial Mail*, 21 September 1990).

In the United States the giant chemical company, Du Pont, is by far the largest emitter of toxic chemicals into air, water and the soil (*The New York Times*, 14 October 1990). In cleaning up its act, Du Pont has built waste water purification plants in which microorganisms digest organic chemicals. Building upon expertise gained in cleaning up its own plants, Du Pont established a safety and environmental resources division and is now selling its services to others (*Fortune*, 1990).

A green movement is essential — as long as it is a green movement that works alongside the trade unions and other organs of civil society

The electrification of Soweto will be both the solution to its hideous levels of air pollution and a major investment which will create new jobs and new demands for electrical appliances. Similarly, improved health standards can cut medical care costs.

Natal's rivers are facing a massive sewage pollution threat: a lack of clean water and sanitation in Durban's sprawling squatter settlements is a growing problem. It is thus interesting to learn that Ulster University is buying all the sewage it can get to provide methane to produce electricity for all the university's needs. It saves energy costs and can help arrest further accumulation of greenhouse gases (*The Sunday Star*, 4 November 1990).

In seeking a growth path based on growth through redistribution, employment and growth will be achieved by addressing basic needs in our society. The alternative path presented by many — particularly big business — is a growth first, redistribution later path. In terms of this vision, growth and addressing basic needs represent conflicting interests.

This supposed conflict is similar to the clash of interests business often sees between pollution control in industry and loss of jobs. As we have just argued, the growth path we seek out must be one where pollution control creates employment instead of destroying it.

Civil society and collective rights

Earlier we mentioned our third reason for saying that red and green should not diverge in the context of a future growth path, namely, to strengthen civil society. How will the environmental impact of economic and political decisions be taken into account? The answer to this is ideally located between two possible polarized extremes: either the state is responsible for protecting the environment, or we allow market forces to protect the environment, with industries either taking upon themselves or internalizing the costs of keeping pollution down.

Experience shows that neither of these options can be relied upon. Even if we have a democratically-elected government, the state bureaucracy needed to police the relevant laws and regulations would be too expensive to sustain. The inevitable response of governments faced with this dilemma is to increase the number of regulations. This only exacerbates the problem — there is a regulatory 'noise' in the economy, but no effective action is taken.

The central planning system of the Soviet economy showed itself to be grossly inept at protecting the environment. Yet the 'free' market has been as bad. Each producer takes responsibility for its own costs (internal costs), but cannot easily take responsibility for the costs of its actions as felt by others (external costs). In relation to the environment this is a particularly serious problem, because by the time consumer resistance starts to force producers to internalize pollution costs, the damage may have been done.

As we mentioned earlier, there are between 5 and 6 million known chemical substances and this number is increasing by approximately 1 000 each year. Only 22 per cent have been studied for toxicity, while 11 per cent have been partially researched. Less than 1 per cent have been completely evaluated (Frumin, Director of Occupational Safety and Health, Amalgamated Clothing and Textile Workers Union, United States, personal communication 1990). We are flying blind and hoping that we will not crash into some unknown chemically-induced disaster.

State intervention in the marketplace to deal with pollution is inevitable. However, the question remains of how this is to be achieved. Before answering this, we need to remind ourselves of the COSATU arguments that there is a need for industrial restructuring and that the unions must take the initiative here. This suggests a critical role for unions: we must not simply depend on the state to carry out the restructuring, but the unions must be allowed to participate in the process.

Our argument for environmental protection is similar. The central state needs to create an effective and coherent framework within which we may utilize our environment and it should facilitate the the exercising of civil society's right to protect that framework. A green movement is essential — as long as it is a green movement that works alongside the trade unions and other organs of civil society.

This conception of civil society is important: civil society acts as a counterbalance to central state power, but in so doing makes the central state more effective and more democratic. Essential to this is the acceptance of collective rights to our economic future and to our environment.

If private property reigns supreme and there is an absolutely free market, then it will not be possible to protect worker rights, nor all our rights to a healthy and smoothly functioning environment. It is increasingly clear from the experience of industrialization — be it capitalist or socialist — that we have to acknowledge collective rights. Not only do we have to acknowledge them, but we have to nurture and facilitate them by allowing people to organize around the protection of these collective rights.

Such organization is the domain of civil society and does not reside in the periodic elections of governments. Democratic elections of the governing party are essential; however, the existence of wider collective rights makes this insufficient on its own. Strong democratic organization in civil society will identify and protect legitimate collective rights. The role of a democratic state is to provide the framework within which such pressure can be brought to bear in a national and international context.

Such considerations are now being intensively debated in the trade unions. There can be no doubt that the green movement is fundamentally based on the acknowledgement of collective rights. It is in the interests of our society that red and green unite as we build a post-apartheid South Africa.

In seeking a growth path based on growth through redistribution, employment and growth will be achieved by addressing basic needs in our society

ESKOM

ABOVE
A coal-fired power station

•HUMPHREY NDABA•

Humphrey Ndaba is walking a tightrope. As a committed unionist and a committed environmentalist, he has to balance the needs of the members of the South African Chemical Workers' Union (SACWU), of which he is general secretary, and those of an environment threatened by the very factories which provide those members with their livelihood. Sometimes the two elements meet when workers' health is threatened by the products they work with.

It is not an easy task — yet he seems to walk confidently on the tightrope and is certain that his dual role is both appropriate and essential.

Ndaba did not set out to be a trade unionist, nor, for that matter, an environmentalist. He was going to be a lawyer. Having graduated with a B Juris from the University of Bophuthatswana in 1984, he joined the legal department of NACTU, where he handled cases of workers who had been dismissed.

His job with SACWU began in 1987 and his interest in health and safety began to grow after he helped collect evidence for lawyers acting against a chemical factory. They were working on behalf of the family of a worker whose death appeared to have been due to asphyxiation caused by nitroglycerine.

'I found that health and safety were inextricably interwoven with questions of the environment,' says Ndaba. Apart from the dangers inside factories, where indifference on the part of management often means that workers are exposed to harmful substances, pollution affects the health of whole communities. In addition, Ndaba points out, many companies processing dangerous chemicals are situated near townships.

As his concern grew, Ndaba became involved in the formation of an environmental workshop: 'through this interaction we developed and now we are involved in community projects.'

A major concern to all South Africa's environmentalists is the question of toxic waste dumping and the temptation of Third World countries in search of revenue to accept the lure of foreign capital in return for putting their people in jeopardy.

'What makes me really concerned,' says Ndaba, 'is that we assume that by 1992 we will have a united Europe; and with the changing global situation, without the conflict between East and West and the waste dumping in Eastern Europe, one sees us as a potential target. How can we minimize or control it?

A microcosm of the toxic waste problem can be found daily in South Africa, where workers will risk their health to work in chemical factories because they are desperate to earn a living. 'It's a situation of jobs versus certain health hazards. The health hazards seem a distant problem when the choice is poverty.'

The problem is aggravated by the fact that the more dangerous industries are inclined to pay higher wages, so union support within the companies tends to be low even when the dangers are explained.

'That trend is worrying. The question of wealth and economics always comes back and flies in our faces,' says Ndaba, giving as an example the readiness of the Namibian government to accept seal culling.

'The poorer you are, the more difficult it is to control the situation. The richer the country becomes, the easier is the battle to strive for a cleaner environment. The situation of high unemployment in most developing countries makes it very difficult.'

The economic realities make trade unions appear conservative on environmental issues but, asserts Ndaba, 'it's not a question of being conservative, unions are more concerned about jobs. The more you have a system which doesn't generate wealth, the poorer you become and the

less careful about environmental issues.

'No matter how wishful you are of a better situation, nature orders that the priority is immediate survival. The programme we have must be generative of wealth, so the task should be easier.'

Thus Ndaba is campaigning both for jobs and for the environment. 'It's quite tricky but it can be done. My worry is, if we are going to interfere with the wealth-generating mechanisms, we can forget about a better environment.

'The system you talk about should be generative of wealth if you want to have a better system and a better environment. The drive for profits negates the move towards a safer planet.'

He maintains that there is growing support among workers, who are increasingly being conscientized about industrial danger, and he would like to see environment groups linking up with the unions in practical campaigns. These could focus on issues such as demands for adequate ventilation and emission systems and pollution monitoring in factories.

Ndaba also believes that environmental organizations must take up the issues of safer substitutes. While pressure might initially lead to redundancies, he concedes, the creation of substitutes could result in whole new industries to which workers can be relocated.

Always it comes back to a question of money. 'The degree of wealth determines the capacity to fight. The battle for a better environment involves money, capacity, improvement. Because this country needs investment, does that necessarily mean we must become a dumping ground for unsafe industry?' he asks.

While he has no quarrel with environmental organizations, Ndaba does see many of them as being engaged in 'moral niceties while we are engaged in the hard issues.' It will be a long time before those issues include the ozone layer or the black rhino. 'We do consider the ozone layer, but it's the last item on the agenda.' He is far more concerned with replacing coal fires with electricity and tarring the roads that are lining people's lungs with dust. Such projects, of course, bring environmental issues squarely into the political arena. The alternative to coal burning is electrification, and the only way electricity can reach all the people is if government policy were changed and electricity were subsidized.

Right now, Ndaba points out, 60 per cent of the people have no electricity. 'It could take a decade to supply everyone with electricity. We need to reduce that decade. We must do something about it. The whole thing has to do with the disbursement of the cake. We are not talking about the end income, but the fringe benefits — right now we have a skewed allocation.

'So, the story of the environment is really the story of poverty trying to improve itself. The more those issues are upfront, the more you get involved, the more the whole subject becomes intricate.'

When he looks at the issues that will face the government that will run tomorrow's South Africa, Ndaba's scenario is an intensely practical one. 'A realistic expectation of a future environment policy would be a time when roads are tarred and the government pushes electrification; a time when there is informal, labour-intensive business so that we can have the choice at the end of the day whether we have to accept people who come with environmentally polluting industry. We must have alternatives.' ❐

7 THE IMBALANCE OF POWER

◆ MARK GANDAR ◆

Energy in all its forms

Good morning, South Africa

On a typical morning in South Africa, nearly half a million people, mostly rural women, set out on the routine but arduous job of collecting firewood. The next day it will be the turn of another half a million or so, and in about four days' time today's gatherers will have to go collecting again. Many will have started in the half-light of predawn, for the task might take up to nine or ten hours and involve walking a total of 20 kilometres to bring home a headload of firewood weighing about 40 kilograms.

They will have been walking for quite some time when the kettles begin boiling and the toasters start popping in suburban homes. *Their* occupants begin their day with the convenience of hot water on tap, of hairdryers and electric shavers. Then they, too, set off on the business of the day, but in a slow-moving procession of cars, most carrying a single occupant.

Apartheid planning does not score well on energy efficiency

Many faces and facets

There are many very different perspectives on the energy-environment relationship. It is viewed differently by the wood-gatherer who notices the recession of trees; by the wealthy professional jealously guarding a highly consumptive lifestyle; by the worker living under a pall of township smoke from wood and coal stoves and commuting absurdly long distances to work (apartheid planning does not score well on energy efficiency); by the environmental activist campaigning for cleaner air or lead-free petrol.

Energy, as a sector, impacts on the environment more than any other. It is associated with a vast range of different environmental problems. The mining and burning of coal and its conversion to oil by SASOL; the use of vehicles for mass transportation; spillages and discharges of crude oil; the handling of nuclear materials and the creation of nuclear waste; the cutting of trees for firewood and the burning of cattle dung; the smoke of domestic fires; the landscapes dominated by transmission lines; the flooding of valleys for

M J UNDERWOOD

LEFT
Rural women carry firewood for fuel against a backdrop of electrical powerlines

Each year roughly eight million tons of wood is burnt as domestic fuel

hydroelectric power — these are by no means all of the energy-related agents and activities which scar and poison the environment. How environmentally friendly or unfriendly is energy in South Africa, and what might be done to improve things?

The poor person's energy crisis

Let us begin with firewood, which provides domestic heat and warmth to over 10 million South Africans and occupies a staggering amount of human time and effort in its collection — two and a half times the total amount of work employed in the country's entire coal mining industry.

Each year roughly 8 million tons of wood is burnt as domestic fuel. It comes from several sources, but mainly from indigenous bushveld or woodlands which cannot yield this amount indefinitely. If no steps are taken to grow more firewood by planting trees the consequences could be dire. Researchers have projected that, if the present trends of wood consumption continue, the 'homelands' could be stripped of nearly all natural woodland by the year 2020. In KwaZulu, 250 forests were proclaimed in terms of the 1936 Land Act: a mere fifty remain intact today. It is unfair to lay the blame for all of this at the door of the wood-gatherers, but firewood is becoming an increasingly dominant factor in the decline of trees. With tree loss the rate of soil erosion, already serious, increases. Ground and surface water hydrology are disrupted and river flow becomes ephemeral, with alternating flooding and desiccation. Dams silt up; fish stocks may be reduced. The nutrient cycle, too, is disrupted, especially if cattle dung is used as a supplementary fuel instead of being allowed to return to the soil.

The amount of plantation required to meet the firewood deficit in South Africa is not trivial — some 500 000 hectares by the turn of the century. In the century that has elapsed between the planting of the first woodlot in 1893 and the present, only 5 per cent of that target has been met.

Woodlots alone will not solve the problem. Another option is a more intimate blend of tree planting and agriculture known as agroforestry. The beauty of this system is its potential for doing environmental good. For example, the judicious use of leguminous trees such as *Leucaena*, which increase the nitrogen content of the soil, may rejuvenate the soil and increase agricultural production as well as provide fuel. Instead of competing with agriculture for land, trees can enhance agriculture. Agroforestry is based on the ecological adage that the whole is greater than the sum of the parts.

South Africa's large commercial forestry sector produces almost enough waste wood in the form of plantation residues and wood-processing waste to meet the demand for firewood in the 'homelands'. It is not uncommon for a resource which is in short supply and severely overtaxed in black rural areas to be wastefully and inefficiently underused in other areas. There it is regarded as waste, or even as a fire hazard, and disposing of it carries both financial and environmental costs.

Fuel-efficient stoves can enable people to use less firewood and can thus slow the rate of deforestation — but the main environmental benefit of a stove is the smokeless environment in the home itself. Soot scrapings from the insides of huts contain carcinogenic substances, and evidence comes from East Africa of a correlation between open indoor fires and the incidence of nasopharyngeal cancer.

Firewood is slowly giving way to other types of energy. In peri-urban areas, transitional fuels like coal, paraffin and bottled gas are being used increasingly, though so far these have had little effect on rural firewood demand. In the electricity supply industry we find another of the sad ironies of South Africa: although ESKOM generates 60 per cent of all the electricity of the African continent, 70 per cent of South Africans do not have electricity in their homes. Efforts are now being made to speed up the electrification of the less developed rural areas, but it is clear that wood will remain the main source of energy in these areas for the foreseeable future. ESKOM's existing power stations are capable of producing about 50 per cent more electricity than they are being asked to at the moment, so extending the grid will not strain the system. This leads us onto the subject of electricity generation and the problems which surround it.

Coal: black gold or environmental scourge?

Coal is South Africa's main energy source: 80 per cent of primary energy in the country comes from coal. Of this about half is used for electricity, a quarter is burnt directly in the industrial or domestic sectors, and a quarter is converted into SASOL liquid fuel.

Coal is extracted by open cast strip mining, a process during which the earth is literally flayed alive. (Instead of drilling underground shafts, the mining company strips away the overlaying earth in toto). Despite an expenditure on rehabilitation of more than ten times the market value of the land, problems are encountered with subsidence, compaction of soil, and drainage. Important wetland ecosystems are lost. The ground water system is completely disrupted when underground water in the vicinity drains downwards through the mine. Chemical changes in mine-related materials impair the water quality of both ground water and runoff: of particular concern is the action of air and water on chemical compounds in the soil, such as iron pyrites, since this produces sulphuric acid. Unsightly heaps of coal discards, some of which have been smouldering uncontrollably for decades, also add to acid pollution. Acidity in water, undesirable in itself, may cause further chemical reactions and release dissolved minerals into the water.

In South Africa the strategy is to try to make the pollution go somewhere else, instead of trying to reduce it — it's cheaper that way

The environmental implications of mining are covered in law mainly by the Mines and Works Act, although numerous regulations in other acts also apply. Theoretically, several ministries are involved. The Government Mining Engineer is responsible for the enforcement of the Mines and Works Act, so environmental control effectively falls within the mining establishment. Enforcement is not strict and penalties are small: in fact, major mining companies took the lead with self-imposed controls, and legislation followed belatedly. Environmental requirements with regard to strip mining are not as stringent as those which apply in the United States, for example.

The most notorious impacts of a coal-based energy economy are those associated with burning it. The half-million or so tons of smoke particles which South African power stations spew into the atmosphere each year are accompanied by about four times as much acid-forming sulphur and nitrogen oxides, as well as smaller quantities of other noxious gases. The eastern Transvaal, the powerhouse of South Africa, has atmospheric conditions which are very adverse for dispersing pollutants. Very little monitoring has been done of air pollution in South Africa, and what has been done has mostly gone unreported. Only recently has an attempt been made to assess air pollution in the eastern Transvaal, and that study by the Council for Scientific and Industrial Research (CSIR), concluded that the emission of sulphur and nitrogen oxides in the area 'is in the same league as the major industrial areas of the world'. The acidity of the rainfall, too, can match that of Europe and North America, where its effects on forests and freshwater life are well known.

Air pollution levels such as those commonly experienced in the eastern Transvaal have been shown in experiments to depress crop growth and impair soil fertility. There is little question that atmospheric pollution has had adverse effects on forestry plantations in Germany, Poland and Czechoslovakia. At least half of South Africa's plantations as well as thousands of hectares in Swaziland are in areas potentially affected by pollution from coal-fired power stations. A study of the Ohio River Basin in the USA by the Institute for Ecology estimated that corn production was reduced by 11 per cent, soya bean yield by 24 per cent to 41 per cent and forest growth by 20 per cent as a result of air pollution related to energy.

In South Africa we do not, as yet, have the spectacular symptoms of air pollution like dying forests and lifeless lakes. We have only circumstantial evidence that air pollution reduces crop production. Also, a recent marked drop in the fertility of breeding cattle in the eastern Transvaal Highveld is alleged to be the result of the displacement of an essential trace element, selenium, by sulphurous rain.

Air pollution levels such as those commonly experienced in the eastern Transvaal have been shown in experiments to depress crop growth and impair soil fertility

The effect of pollution on health is difficult to assess: health is so complicated and multifaceted that it is almost impossible to evaluate one factor in isolation. There is some evidence, albeit inconsistent, to link slight increases in respiratory ailments in children to air pollution in South Africa. But what of the rest of the public and what of synergy with other factors? The aged, for example, or emphysema sufferers or heavy smokers, are more at risk than healthy people. Taking such factors into account, the Brookhaven National Laboratory in the United States estimated that in 1979, air pollution from coal-fired power stations caused or contributed to some seventy-four deaths per 1 000 MW of generated power. This is probably the highest conceivable estimate since Brookhaven operates under contract to the nuclear industry — but if this is correct and the same sort of figure applies to the eastern Transvaal, then the number of pollution-related deaths there would be in the order of thousands per year.

So what are we doing about it? In South Africa the strategy is to try to make the pollution go somewhere else, instead of trying to reduce it: it's cheaper that way. The newest power stations have 300 metre tall stacks to carry the gases above the inversion layer. This does indeed reduce air pollution in the vicinity of the power stations, but instead of being completely dispersed, the pollutants tend to be trapped in an atmospheric belt. As a result, an even larger area is susceptible to pollution and instances of severe air pollution may occur sporadically as a result of this accumulation.

Studies in the United Kingdom have shown that lethal cancers in populations living near nuclear installations are two to ten times more common than the national average

Flue gas scrubbers, which are standard on power stations in the United States, are not used in South Africa. These can reduce sulphur dioxide emissions by 50 to 90 per cent, albeit with some loss of efficiency. However, there is now a new breed of power station. Coal was traditionally burnt in the form of lumps, but these new stations use a simple process to turn coal into a combustible gas. This coal gasification process makes a further tenfold reduction in sulphur dioxide possible and is accompanied by greatly reduced emissions of the oxides of nitrogen, and an increase in overall efficiency.

The rate of emission of carbon dioxide, one of the greenhouse gases, is less responsive to technology. More efficient technologies produce less carbon dioxide per unit of electricity, but the improvement is small compared to the 50 to 80 per cent reduction which may be necessary to avoid major climatic change.

Some newly developed technologies generate electricity from coal more efficiently and very much more cleanly than conventional power stations. The rate of emission is expressed as a percentage of that from a conventional power station with scrubbers. Without scrubbers SO_2 emissions would be nearly ten times greater.

	Per unit of electrical energy			Efficiency	Capital cost
	SO_2	NO_x	CO_2	%	$ per kW
Conventional with scrubbers	100	100	100	34	1 600
Gasification, combined cycle	10	8	80	42	1 700
ISTIG with coal	10	7	80	42	1 030
ISTIG with natural gas	n	3	40	47	400

n = negligible amount
ISTIG = intercooled steam-injected gas turbine
(From W Fulkerson, R Judkins and M Sanglivi in *Scientific American*, September, 1990)

Garritt Hardin coined the phrase 'the tragedy of the commons' to describe the inevitable destruction of common resources if left to a free-for-all. The benefits of overexploitation (short term profits) accrue to individual exploiters, while the costs of destruction are shared equally by everyone. The atmosphere is a global commonage and the greenhouse effect is a tragedy of the atmospheric commons. The situation requires unprecedented international cooperation, and it is hard to feel optimistic about the likelihood of this happening in the near future. At least we can help in the interim by using coal more slowly in more efficient power stations. Greater reductions of carbon dioxide per unit of electricity could be achieved by using natural gas for electricity, but South Africa's gas is already committed to liquid fuel production.

South Africa is not developing any of these new technologies for various reasons. They are not well suited to the colossal 4 000 MW power stations currently in vogue here: with these technologies we would have more and smaller stations. Our electricity might be a little less cheap than at present, a point I shall return to later. But perhaps the main reason is a preoccupation with the idea that the days of coal are numbered. Soon there will be none left, so what's the point of investing in the development of new technologies? The last coal-fired power stations, we are told, will be built in the first half of the next century.

Thermal and fast-breeder reactors

Uranium is not simply uranium: there are two different kinds, or isotopes. The type which splits easily and gives off energy is designated 235. Less than 1 per cent of natural uranium is 235. Thermal reactors derive their energy from uranium 235. In fast-breeder reactors, the apparently useless part of uranium, 238, is converted into plutonium. This is the stuff that nuclear bombs are made of, but it can also be used as a fuel. Fast-breeder reactors activate uranium 238 faster than they use up their fuel — in other words they breed nuclear fuel.

If South Africa's uranium were used in thermal reactors only it could generate only 25 per cent of the energy available in our fossil fuel reserves. Globally the figure is a mere 10 per cent. Without fast-breeder reactors, the nucleus will not be a source of abundant energy. But fast-breeder reactors have thus far proved to be rather temperamental.

But need we in fact exhaust our coal by then? At present, power stations use barely one thousandth of the country's coal reserves each year. If present rates of extraction of coal are maintained, our known coal reserves will see us into the twenty-fourth century. It is not essential for South Africa to mine the coal faster for the time being. A quarter of this coal is exported and a quarter of the remainder feeds the SASOL coal-to-oil process. This process is environmentally messy and energy-inefficient; it was adopted here in response to the oil embargo against apartheid. No other nation since Germany in World War II has seriously considered this process. Furthermore, there is much that can be done to eliminate inefficiency and wastefulness in energy consumption. If we want to use our coal to fuel relatively clean-burning power stations for many years to come, we have the option of doing so.

Nuclear power

Nuclear power has been vaunted as a clean and abundant alternative to coal. In the late 1960s and early 1970s, the South African authorities anticipated that there would be several nuclear power stations dotted along the coastline by the end of this century. As it turned out, energy demand grew more slowly than had been expected and a nervousness about the economics of nuclear power crept in, so South Africa will have only one nuclear power plant in the year 2000. However, although the time frame has been put back, the old gung-ho attitude to nuclear power persists. This was the vision expressed by the chairman of the Atomic Energy

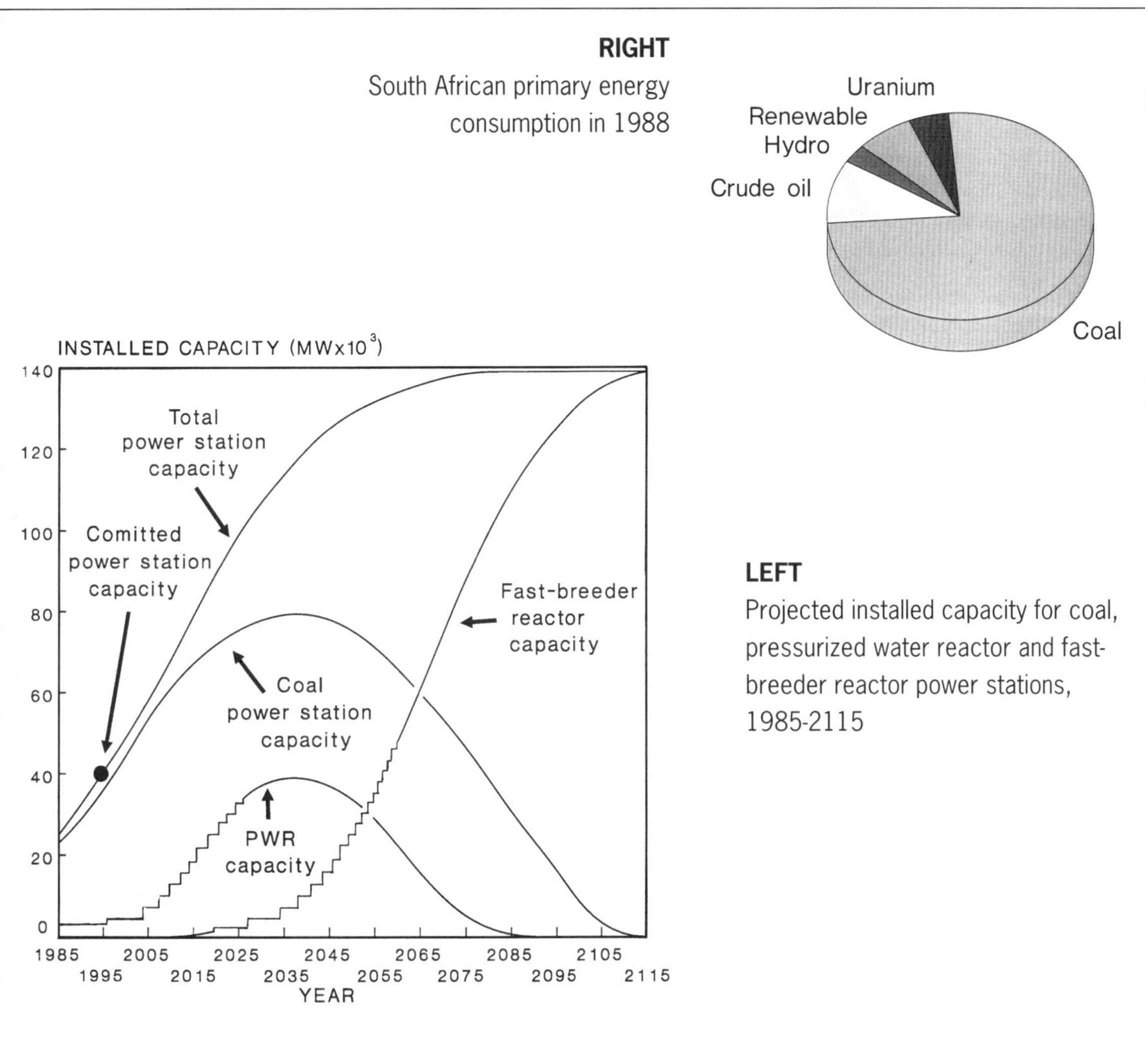

RIGHT
South African primary energy consumption in 1988

LEFT
Projected installed capacity for coal, pressurized water reactor and fast-breeder reactor power stations, 1985-2115

Corporation in 1985: 'coal will still be our main resource for electricity generation until approximately 2050. Thermal reactors will provide an increasing share from 2000 onwards, peaking at an approximate share of 50 per cent around 2060. Fast-breeder reactors will then have to take over from coal and thermal reactors.' (See the graph above for official projections of the role of nuclear power in the future.)

Accordingly, ESKOM has undertaken a major search for sites suitable for nuclear power stations. So far three have been pinpointed and the land is being acquired. It has been said that these sites are merely to be put aside for possible future nuclear development, but other pronouncements suggest that there are more immediate plans afoot. A recent statement by the chairman of ESKOM indicates the commissioning of a new nuclear power plant by 1995, with construction starting before 1998. Thereafter at least one nuclear power station unit will be built every few years.

Nuclear power introduces a new set of environmental concerns related to radioactivity. Nuclear power carries with it a host of associated nuclear facilities, making up the nuclear fuel chain which starts with uranium mining and ends with the disposal of long-life radioactive waste. At each link in the chain workers, the public and the environment are at risk from routine and accidental releases of radioactivity. The effect of these small but insidious exposures to radiation is a matter of controversy. If someone develops cancer, it is hard to pinpoint exactly what caused it, and obviously it is not possible to do controlled experiments on humans. At best, evidence is circumstantial.

A recent article in *New Scientist* suggests that the widely-accepted estimates of the effects of radiation on humans have been based on faulty assumptions and that harm could be caused by smaller doses than was previously thought, maybe even by a factor of ten.

Studies in the United Kingdom have regularly shown that lethal cancers in populations living near nuclear installations are two to ten times more common than the national average, even if there has been no major reported leakage of radioactivity. These statistics, though impressive, do not necessarily imply very large numbers of deaths, since the baseline incidence of these cancers is low. Although there is only one nuclear power station in South Africa, there are a number of other associated nuclear installations, but it has not been established whether or not these are ringed by areas of increased cancer risk.

Uncertainty also surrounds the possible effects of a major accident at a nuclear plant. In the aftermath of the accident at the Chernobyl nuclear power station in the Soviet Union, several 'authoritative' estimates of the number of resulting cancer deaths over the next twenty years were published. These varied from 'a mere handful' to 750 000 over two continents. We shall never know the final count. At first 116 000 people were evacuated from the vicinity of the power station. Three years after the event a further 100 000 were moved after a rethink about the nature of radioactive contamination.

Now, five years afterwards, Soviet scientists admit that people in the region are still eating and breathing radioactivity. One and a half million people there are said to have suffered effects including radiation sickness, fibrosis, thyroid cancer and leukaemia. It is planned to move up to a million more people to enlarge the radius of the depopulated hole in the Russian map from 30 to 100 kilometres. In this time of thawed international relations, South Africa has offered assistance with their relocation. That nuclear cooperation between the two countries might begin in this way is not without irony.

ESKOM

ABOVE
Children walk on the beach on the West Coast with Koeberg power station under construction in the background

Asking the wrong questions

South Africa's energy economy is already tilted heavily towards electricity. Some projections envisage an even stronger dependence on electricity and a four or fivefold increase in electricity generation over the next hundred years or so. Embodied in these projections is

the premise that only coal and nuclear power have any significant contribution to make. From an environmental perspective, we are locked into a gloomy and sterile debate about which is the worse of two very considerable evils. It is gloomy because of the enormity of the possible effects, and sterile because these are so different. One can compare apples and oranges for ages without getting anywhere. Perhaps we have been asking the wrong questions.

It is not the purpose of this article to catalogue all energy-environment relationships from petrol engines to harnessing ocean wave energy. Let us rather turn attention to the assumptions and attitudes which underpin the prevailing energy situation in South

Unusually high incidences of cancer have been found to be associated with nuclear installations in the United Kingdom.

SELLAFIELD reactor and reprocessing plant in Cumbria: child cancers and leukaemia in the population near the plant are ten times the national average.

DOUNREAY reactor in north Scotland : leukaemia among under twenty-fives within seven miles of the plant is ten times the national average.

HUNTERSTON nuclear power plant in western Scotland : leukaemia within ten miles of the plant is double the average.

CHAPEL CROSS nuclear power plant in west Scotland : myeloid leukaemia among children in the vicinity is three times that expected.

WINFRITH nuclear power station in Dorset : childhood leukaemias in the vicinity are ten times the national average.

BURGHFIELD nuclear bomb factory in Berkshire : leukaemia in children under five years in the district is five times the national average, increasing to ten times closer to the factory.

HOLY LOCH nuclear submarine base in west Scotland : cancer deaths in nearby villages are three times the average and there are abnormally high rates of deformities among the children of workers at the base.

SIZEWELL nuclear power plant in England : workers have eight times more cases of leukaemia than the average.

Africa. Solutions are to be found in new perspectives on the issue, rather than simply in technological choices.

The energy-growth fallacy

If the Gross Domestic Product (GDP) of various countries is compared with their total energy consumption, there appears to be a close correlation between the two. However, there are some examples which depart from the norm. South Africa is one of these, with a much higher rate of energy consumption per capita or per unit of net output than countries at a similar level of development. Also, within the economic history of an individual country, energy consumption and GNP are not always in step with one another.

There seems little justification for postulating a simple causal relationship between energy consumption and development. Yet in South Africa the belief seems to have prevailed that pumping the economy with energy leads in some magical way to growth, job creation, wealth and economic wellbeing.

In a 1977 report, the Department of Planning and the Environment said, 'the importance of energy to the national economy is much more than its direct contribution to the national product, as it serves not only as a basic input, but also acts as a catalyst for growth'. In a subsequent document in 1979, 'Outlook for Energy in South Africa', it explained that this energy-induced growth 'is necessary for the South African economy in order to provide employment for a rapidly expanding population and to increase the welfare and standard of living of all the people of the Republic'.

As we enter the 1990s, industrial utopian thinking is increasingly being challenged. Growth per se will not provide the number of jobs required; it will not necessarily lead to a contented and stable society; it will not generate the surplus wealth needed to clean up the environmental mess created in the process.

Let us look at the specific question of energy and growth. A useful index of how efficiently energy is being used is energy intensity, that is, the amount of energy consumed per unit of GDP produced. This can be calculated for the economy as a whole or for different sectors. In the South African industrial sector the energy intensity rose by 48 per cent between 1971 and 1986. That means that in 1986 a unit of energy produced much less than it had fifteen years previously. During the same period, however, the energy used in industrial processes in the United States declined by 1 per cent per year despite a 2 per cent per year growth in manufactured output. In mining in South Africa, energy intensity has doubled in the same period, reflecting a switch from labour to capital. In agriculture, too, there has been a steady rise in energy intensity and a corresponding increase in the number and size of rural settlements housing former farm labourers who have been displaced. More energy consumption does not necessarily result in proportionate growth and more jobs.

While the theory that growth follows automatically from energy by some catalytic process is fallacious, all good fallacies contain elements of truth. Energy *is* a prerequisite for the development process, but it must be present in the right amounts, in the right form, in the right circumstances, at the right time and in combination with the other ingredients required for development. In short, energy cannot be planned in isolation.

> **The poor persons's energy crisis is still generally regarded as a separate issue — something outside the serious business of national energy**

From forecasting to force-feeding

The econometric equations used in forecasting energy demand have been worked out in the light of past trends: thus an increase in energy demand in one period to some extent results in an increase in the energy demand predicted for the next. Furthermore, electricity-demand forecasting tends to be self-fulfilling. Supplies are increased to meet predicted increases in demand far in the future because of the long commissioning and construction time needed for large power stations. Thereafter, the power must be sold to recover the high capital costs incurred in setting up the new stations, even if this means stimulating a flagging demand. Thus the rate of increase of electricity demand has an inherent inertial resistance to change.

South Africa has cheap electrical power, and it is even cheaper per unit if larger quantities are used. This is hardly likely to encourage consumers to use electricity in moderation. In addition, the fact that electricity sales are seen as a source of revenue for municipalities adds to the vested interests stacked against conservation. The economy may find that it is being force-fed with electricity.

Forgetting the poor

Since the poor are least able to contribute to the GDP and because they are least able to pay for commercial energy, they have historically been ignored in energy planning and have been left to gather their traditional fuels. Until recently, firewood was not even considered worth mentioning in official statistics on energy in South Africa. The poor persons's energy crisis with which we began this discussion is still generally regarded as a separate issue — something outside the serious business of national energy.

However, there are some encouraging signs that this may change : the Development Bank of Southern Africa supports rural energy projects, as did the National Energy Council until it was disbanded in April 1991; ESKOM is looking at innovative approaches to rural electrification; and there is a growing awareness in certain sections of the forestry profession of the need for new and imaginative ways of producing firewood.

Broadening the energy base

No single prescription exists for a solution to the energy problem. However, if the focus is shifted onto how energy is being used and how it should be used, as well as onto genuine needs rather than gross energy demands, then other possibilities open up. One of these is developing a broad energy base which includes several renewable sources rather than a few non-renewable sources.

Renewable energy covers a wide range of energy sources and energy technologies. These, of course, have environmental impacts which are sometimes minor and sometimes not so minor. The energy option which maximizes the use of renewable sources and incorporates energy conservation, the use of waste products and recycling, is referred to as the 'soft' energy pathway. Overall, this creates more jobs and far less environmental damage than 'hard' energy.

There is a respectable amount of energy to be had from renewable sources in South Africa. According to one projection, the Natal/KwaZulu region could produce all of its electricity and nearly all of its liquid fuels from renewables alone. This could be done using energy derived from wind, the sun, water, sugar, wood residues and agricultural and urban waste. Indeed, even in that region, which has relatively unfavourable conditions for solar energy, solar installations covering only 0,1 per cent of the land area could produce all the electricity the region uses. Even urban areas, using roof-mounted solar collectors, waste products and other local resources could go quite far towards energy self-sufficiency. For example, Durban could produce an amount of energy equivalent to 40 per cent of its electricity consumption if good use were made of these opportunities.

High capital cost has been a barrier to the use of renewable energy technologies on a large scale, but with research, development and large-volume production, some renewable technologies are becoming more and more cost-competitive. To some extent the relative costs of renewable technologies reflect how far research here has lagged behind research on non-renewable energy. Further, the environmental costs of non-renewable energy are not reflected in its price.

Most renewable sources of energy are diffuse and dispersed. They are not suited to large energy installations and have mostly been confined to small, specific niches such as wind pumps and solar water heaters. If soft energy options are to be widely adopted, there will have to be more local autonomy in energy matters to allow local solutions to be found to local needs.

Energy integration in southern Africa

Not all renewable sources are diffuse and dispersed: the exception to the rule is large-scale hydroelectric power. The hydroelectric resources in South Africa are small compared to the huge untapped resources in Zaïre, Zambia, Zimbabwe and Mozambique. It has been estimated that Zaïre alone could supply all the electricity requirements of the subcontinent until the end of the century. Furthermore, there are large coal and oil deposits within the Southern African Development Coordinating Conference (SADCC) region.

Through SADCC, the possibility of regional energy partnerships is being explored and there is a very real possibility that a post-

RIGHT
Phola Park squatter camp near Thokoza township on the East Rand

JUSTIN SHOLK

apartheid South Africa would join the SADCC community. This could have major implications for the future energy scenarios for the region.

Large-scale hydroelectric power does have considerable environmental and social costs. Hydroelectric schemes in the Third World are drawing increasing criticism from environmental quarters. Flooding causes substantial losses in terms of wilderness areas, agricultural land, archaeological sites and so on. Communities are displaced and often impoverished, social cohesion breaks down, the incidence of disease increases and secondary environmental damage occurs in the resettlement areas.

The construction activities themselves cause great damage to remote areas, accompanied by roads, quarries and transient boomtown effects. The water impoundments themselves trap silt, and may disrupt the ecology of the lower river. They provide breeding grounds for carriers of diseases and for water weeds.

One of the main criticisms of Third World hydroelectric development is of their political dimension. Hydroelectric schemes are said to be built mainly for the political prestige of a ruling elite and at the expense of the rural poor. In southern Africa there could be further political and ethical issues: South Africa, a country which already generates 60 per cent of Africa's electricity, would be drawing hydroelectric power from less developed neighbours while they were left with the environmental and social consequences.

Energy conservation and efficiency

As we saw from the energy intensities mentioned earlier, there is much unproductive and wasteful energy use in South Africa. There is little incentive to conserve energy.

However, energy consumption can be reduced in any number of ways. It can result from improved methods of production in industry, and from the manufacture of more efficient appliances. Energy consumption can be reduced in transportation by eliminating the permit system, which is responsible for the absurd 'empty leg' in the road transport of goods (trucks carry a load to its destination and return empty); by improving traffic flows; by encouraging mass transport systems; and by allowing people to live near their places of work. It can be reduced in housing and in the heating and cooling of large buildings.

Energy is thrown away in the waste products of agriculture and forestry and in urban refuse and sewage. Recycling can save a portion of the energy invested in the initial manufacture of the product, particularly in the cases of paper, glass, steel and aluminium. But energy conservation goes deeper than this.

It is part of resource efficiency in society; it implies the optimum use of resources for all members of society. It is only possible to have true energy conservation in a resource-conscious society.

BELOW
A working wind farm in Palin Springs, California

Conclusions

Energy comes from different sources, in different forms, via different technologies. All together, energy as a 'sector' produces an array of environmental impacts of bewildering variety and often alarming seriousness. Much can be done in the short and medium term to alleviate harmful environmental impacts, to develop a broader and more environmentally friendly energy base, and to encourage conservation and energy efficiency and eliminate wastefulness.

Some of the environmental effects are rooted in socioeconomic factors such as poverty. There is an urgent need to redress historical inequalities, and to include the requirements of the rural and urban poor in energy planning.

There is also a pressing need to find new perspectives on the relationship between energy, development and the environment. We need to re-evaluate many economic assumptions and look critically at what the economy produces, and how it does so. In the long term, the sustainability of an economy is dependent on the way in which energy and other resources are used.

Solutions are to be found in new perspectives on the energy issue, rather than simply in technological choices

•MARK GANDAR•

There is a certain logic in the fact that the founder of the Society Against Nuclear Energy (SANE) should once have been involved in nuclear physics. Who better, after all, to know the dangers?

Still, if anyone had told Mark Gandar when he began his studies in the 1960s that he would end up taking an anti-nuclear stance, he might not have believed it. His is an odyssey which has irrevocably altered not only his lifestyle, but also his priorities, his philosophy and his response to technology. A practical scientist, he now has a growing belief in a spiritual dimension.

Gandar's relationship with the splitting of atoms began when he worked in the nuclear physics research unit at Wits after he graduated. His career plan was simple then. He would go on to Oxford to continue his studies in physics and would then get involved in low-energy nuclear physics. The problem was that he became 'disillusioned with quite a few things' — Oxford and physics amongst them — so he gave up his studies and went to Botswana instead.

The experience in Botswana did two things for Mark Gandar. It taught him the joy of the outdoors and it opened his eyes to environmental issues. He began to realize that the problems of the developing world were less likely to be solved by economists than by people with a knowledge of the environment.

Determined to be part of the solution, he proceeded to Aberdeen University to do a masters degree in ecology — a word he had not even heard, much less considered a serious subject, during his undergraduate years.

On his return he found himself in the northern Transvaal studying grasshoppers — for four years. His move back into human ecology came when he began to look into questions of firewood and tree removal — the effects of the harvesting of trees in rural areas. He joined the Institute of Natural Resources in Pietermaritzburg and became a member of the Institute of Ecologists, a move that brought him and nuclear physics together again, this time at loggerheads.

An article in the Institute's bulletin presenting both sides of the nuclear power story worried him. 'I had a gut feeling that the proper stance of ecologists on nuclear power was not necessarily a neutral one.'

As he researched information for the letter he intended to write on the subject, Gandar became aware of the environmental aspects of nuclear power.

'I delved into the subject and became so fascinated I got quite heated.' The result was an invitation to give a paper at a seminar on nuclear power. 'It was produced as an anti-nuclear input at a largely pro-nuclear seminar.'

Involvement grew, and in 1983 Gandar started SANE — an organization, he observes, which brought together quite a motley selection of people, most of whom had only one thing in common — their opposition to nuclear power.

They were drawn from groups as divergent as the wildlife societies, the End Conscription Campaign, the peace movement, feminist organizations and Rape Crisis, and they all brought their particular perspectives to bear upon the issue.

'The result for me was that I developed much more of a philosophy around nuclear power and environmental issues and saw nuclear power not simply as a technical and environmental issue but as political and philosophical.'

Initially the issues were the obvious ones — opposition to the search for more nuclear power sites; the spillage of uranium in an accident on a Natal road; the non-signing of the nuclear non-proliferation treaty. Later, though, the ramifications of the issue led to a questioning of some of the fundamental beliefs of a technological society

and through that to a search for alternatives.

'We try to bring about a sort of understanding not only of the danger of nuclear power but, in a political sense, the nature of a nuclear state; the degree of centralization of power; the secrecy; the restriction of information; the belief in — the almost worship of — technology; unquestioning attitudes to the idea of endless growth . . . the whole ethos,' says Gandar.

He concedes that there are aspects of nuclear energy which have useful and positive benefits — for instance, the use of radio isotopes in medicine and in research. 'But I don't see any way in which the good can be separated from the bad. You can't have the spinoffs without the basic nuclear industry — and it is an industry, from the mining of the raw materials to the processing, use, clearing up of the mess, storage, and waste.'

He does not claim, either, to know what the solutions are. 'I don't have a really nice vision of how we get in ten easy (or difficult) steps from where we are to a completely non-nuclear world but there have been some quite positive signs in certain developed countries which have either abandoned nuclear power altogether or at least put a moratorium on it.'

What he does have is a clear idea of how to begin a campaign. 'Step one is trying to get across awareness of all the implications of choosing nuclear power, which includes the dangers inherent in NP itself. It includes the difficulties and the importance of controlling the spread of nuclear weapons in a world in which nuclear technology is freely exchanged for ostensibly peaceful reasons. It also includes the danger of creating a global economy of radioactive products, particularly plutonium.

'The only way in which one can move towards a non-nuclear world is by somehow forcing the decision upon governments, and the two most powerful forces for changing government policy are economic forces and, if it's really strong, public opinion.

South Africa, says Gandar, is a long way from having a well-organized anti-nuclear movement with effective political clout. Still, he believes that SANE's activities have helped make the issues known. 'I think we have gone a small way towards demythologizing the nuclear issue; that it's something that can be talked about quite openly without being branded as anti-progress or anti-South African. Public interest in nuclear power has forced ESKOM at least to be less secretive in their nuclear developments, and that's a start.'

The issue of nuclear energy and the new technology has, for Mark Gandar, gone far beyond pure pragmatism, pure politics or pure protest. It is a philosophy of life, a dream of a society in which people will not dominate people and people will not dominate the environment.

'I look towards a stage where human values become more important than technological values and people's lives are much less governed by technology. It's a question of going forward to a state like that rather than backward.' ❐

8 FLOWERS IN THE DESERT

Community struggles in Namaqualand

◆ DAVID FIG ◆

In Namaqualand the miracle of spring is particularly evident. Usually an arid region of coastal sandveld and copper-coloured mountains, in winter this land is transformed by the rains: by spring the entire veld seems to be in bloom. Thousands of tourists visit the area to observe the multicoloured floral vistas adorning the earth that is seemingly bare for the rest of the year.

Namaqualand has become synonymous with its startling and fragile ecology. Yet the popular image of the area as a floral wonderland is a politically neutral one that denies the complex social and economic relations in the region.

Because of its remoteness, most South Africans will never discover this part of the country. A peripheral area, wedged between the south-western Cape and Namibia, its boundaries are the Olifants and Orange Rivers, the Atlantic Ocean to the west, and Bushmanland in the interior.

Although it is the same size as the Netherlands, which supports 15 million people, Namaqualand has a population of only 60 000. The low population density that results from the sparseness of agriculture and lack of economic development, means that the people of the region are often disregarded in official calculations as well as in policy formulation.

The climate and the austere terrain of Namaqualand are not its only harsh characteristics: the history of the region is one of extreme poverty and exploitation, of land seizure and violation of ancestral rights, of racial discrimination and forced population removals.

When we speak of the environment, its protection, or its sustainability, we cannot see it in isolation from the communities which it encompasses. When examining rural areas such as Namaqualand, one sees this essential relationship more clearly. The days of forced population removals are apparently over: owing to mass resistance and community organization, it is no longer politically feasible to detach people from their environments against their will. They are there to stay, regardless of the rigours of their environment. Therefore a close examination of social and political issues in an area is a key to understanding environmental questions. Through studying community struggles over land rights, we can begin to see that the concern for a sustainable ecosystem is not only the province of an informed urban middle class.

This chapter hopes to show, through a case study of Namaqualand, that popular rural political struggles of the 1980s and 1990s are largely to do with the land, and hence the use of the environment. Isolated and marginalized for many years, the people of Namaqualand have seen their communal land progressively alienated, restricted, encroached upon, eroded, and privatized.

Trekboere

From the late eighteenth century onwards, white cattle farmers known as trekboere encroached on Namaqualand, moving northwards and occupying land through conquest. Indigenous Nama herders were driven into more arid areas. Towards the mid-nineteenth century, they sought the protection of missionaries, who helped them obtain title to some land located around the missions.

Title took the form of 'tickets of occupation', apocryphally

BELOW
A donkey cart carrying water approaches Kamassies near the Vaalputs nuclear waste dump in Namaqualand

PAUL GRENDON

The concern for a sustainable ecosysem is not only the province of an informed urban middle class

issued by Die Oumies-Koningin (Old Mrs Queen Victoria), granting land for the exclusive use of local people. This land was transformed into small reserves which permitted the communal grazing of livestock. The white farmers were prevented by law from encroaching on this land. However, these laws were not easy to implement, and the farmers failed to abide by them. These reserves still exist, and today one out of every three Namaqualanders lives on a reserve.

The official boundaries of the reserves are contested by the Namaqualanders, who claim that historically they extended over a wider area. Archival research needs to be undertaken to assist the communities of the reserves in asserting their claims.

Diamonds, copper and capital

The next assault from outside the region was by the mining companies. Since the first expedition by governor Simon van der Stel to the copper mountains around Springbok in 1685, Namaqualand was known to be rich in minerals. Copper has been mined for over a century and a half. Since their discovery in 1926, diamonds too have been mined along the mouth of the Orange river and along the Atlantic coast. Mineral wealth in the area also includes deposits of zinc, gypsum, and semi-precious gemstones.

Mining became the dominant industry in the area, alienating some of the reserve land. In the case of diamonds, mining companies gained exclusive control over areas, including the power to determine who might use certain public roads. The land used for diamond mines is largely no longer viable for other purposes. When mining is over, the companies do not return the land to the people: instead, in some cases, they have started their own farming operations.

The mining sector drew most of its labour from the reserves. The reserves began to resemble classic bantustans, with labour migration, a poor infrastructure, high dependency on low-wage incomes and few other employment options. Those who remained outside the mining sector derived income from stockfarming - largely sheep and goats - in the reserves. Many of the mine workers also retained the link with pastoralism, keeping their herds as a form of social security owing to the economically unreliable nature of the mining sector.

BELOW
A map of the Namaqualand region

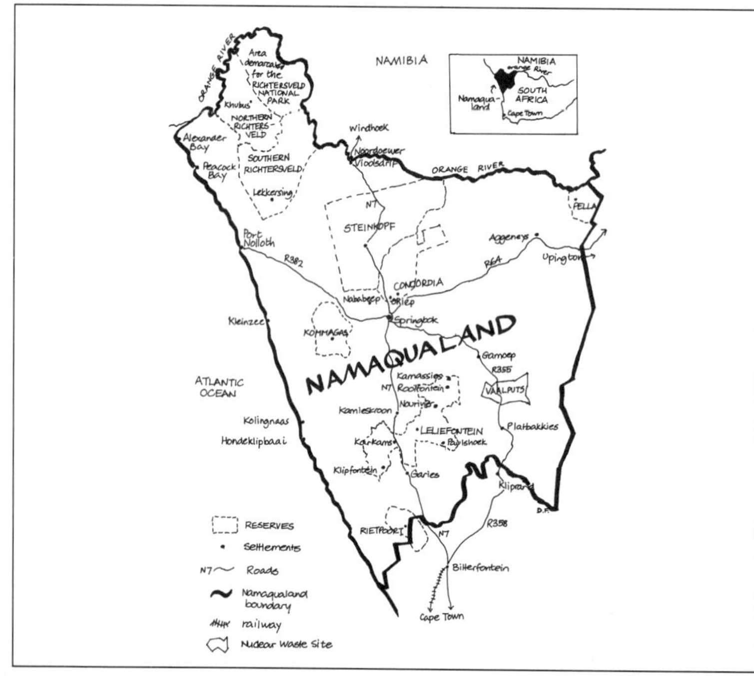

The copper-mining sector is very volatile. The copper mines are dependent on strong world copper prices and good exchange rates to make them viable. Their resources are finite, and closure of some mines in the 1970s had a devastating impact on household incomes in the reserves. As it is, employment has decreased notably on the mines, with significant retrenchment taking place every time there is a drop in the world copper price. The copper mines have remained open only through government intervention.

The diamond industry, despite fluctuations in production, has retained a reasonably stable workforce of around 6 000, whereas employment in the copper mines has declined from a peak of 5 000 in the 1970s to around 2 500 today.

Other sources of employment in the region are the service industries, agriculture and fishing. Namaqualand has no rail system, so that the road haulage industry (dominated by one company, Jowells) is an important sector based on servicing the mines. Agriculture accounts for only 10 per cent of the region's production. Together with fishing, it provides only limited seasonal and casual labour opportunities and offers extremely low wages. Average full-time employment on the white-owned farms is about five workers per farm of an average size of 7 000 hectares.

The precarious nature of employment in the region has heightened the importance of access to the land. In times of recession and joblessness, the people in the reserves can turn to the land to provide minimal subsistence.

Social conditions

Why was mining seen as an assault on the people of the region? Although it provided incomes for mining families, the mining industry transformed the reserves from relatively independent subsistence economic units into reserves of cheap labour. The industry made little attempt to assist in the development of the reserves or the education of the inhabitants' children. In studies produced for the Second Carnegie Inquiry into Poverty and Development, the high levels of unemployment and poverty in the reserves were stressed (Boonzaier, 1984; Sharp, 1984; Sharp and West, 1984).

The mining industry transformed the reserves from relatively independent subsistence economic units into reserves of cheap labour

The incomes of the people living in the reserves are low, and the prices of basic goods are extremely high. Levels of technical education, agricultural extension, health care and other social services are abysmally low. There is a lack of post-primary schooling and clinical health care. The work of Operation Hunger, a welfare organization, has necessarily expanded over the last few years. The degree of malnutrition has led to a significant degree of mental retardation in children (Dunne, 1988).

A dearth of recreational facilities and public transport isolate the inhabitants of the reserves. Most reserves are not linked to the national electricity grid, yet power lines for the mines cross over their territory.

The mines have made little effort to implement social responsibility programmes in any form. The Diamond Fund, administered by the Small Business Development Corporation (SBDC), has gen-

'The law allows private ownership of our inheritance and birthright . . . it deprives us of our rights'

— Mier resident

erated over R11 million in royalties (Thomas, 1990), but inhabitants regularly point out that they never see the fruits of the fund translated into real development. The bulk of the royalties and dividends (64 per cent) is at present allocated to state bodies and not to the community. The SBDC is promoting the establishment of a Trust to administer the diamond fund from 1 January 1991.

While mining provides the bulk of employment in the region, the region's dependency on mining is a dangerous one. The need for sustainable development is not met by the mining industry and when the mines close, as they inevitably must, no alternative employment will become available unless there is better advance planning for sustainable economic activity in the region.

Tricameral privatization

A further assault on the community has come from the apartheid structures themselves. Racial classification of the reserve inhabitants placed them historically under the control of the Department of Coloured Affairs. When the tricameral parliament was established in 1984, jurisdiction over the reserves fell to the Department of Local Government and Agriculture in the (ethnically 'coloured') House of Representatives. Since its inception, this House has been under the control of the Labour Party.

In most of the Namaqualand reserves, the Labour Party has sustained an unpopular and often corrupt local government structure of management boards, consisting of both pliant opportunists and loyal party supporters. Their political positions have allowed members of the boards to allocate resources preferentially to themselves, their friends, and their families at the expense of the community as a whole. Lack of political experience and strong religious belief has inhibited local communities from challenging the authority of the boards until comparatively recently.

Historically, in most reserves, grazing land was communally owned. This enabled all residents to keep flocks in the reserves, to adopt more nomadic and decentralized settlement patterns, and to use a variety of plants for shelter, food, fuel, and medicinal use (Archer, 1990).

Even prior to the establishment of the tricameral parliament, the Labour Party began to apply the principle of privatization to some of the rural reserves of Namaqualand and the Kalahari. Arguing that communal grazing methods were outdated, caused overgrazing, and discouraged entrepreneurship, the Labour Party attempted to split the grazing land into camps (the so-called 'economic units') which would be fenced off and sold to individual farmers, thus foisting a new form of economic differentiation on the community.

Botanists and agronomists challenged the Labour Party's arguments in a study which concluded that 'the so-called "economic farming units" proposed for the Leliefontein reserve are clearly uneconomic and would not have been able to promote development' (Archer, Hoffman and Danckwerts, 1989).

The impact of privatizing the land was to impoverish those farmers left without access to grazing lands. They were forced to sell their flocks to those individuals to whom camps had been allocated. Since the allocation of camps was at best inequitable, and at worst corrupt, the majority of the farmers began to challenge the privatization of land almost as soon as the Labour Party administration began to implement it in the bigger reserves.

At Leliefontein, Steinkopf, and the Southern Richtersveld, challenges eventually took the form of legal action against the management boards and the Labour Party administration. Assistance was obtained from public interest law practices and the Surplus Peoples Project, an organization working to defend communities against loss of land. The legal actions spawned more active political structures in the communities, committed to working against the privatization and challenging the undemocratic practices of the boards.

ABOVE
Mier residents protest against the sale of communal land

During the course of 1988, the Supreme Court of the Cape of Good Hope pronounced in favour of the Leliefontein, the Steinkopf and the Southern Richtersveld communities, and set aside the boards' attempts to lease out individual 'economic units' carved out of previously communal land. The community celebrated their victory with a community feast and cultural events to which supporters from many countries were invited.

The threat of privatization is not really over. Despite the intervention of the deputy president of the ANC, Nelson Mandela, on behalf of the Mier reserve in the western Kalahari, the Labour Party has passed legislation to enable its Minister of Local Government and Agriculture to sell land that is supposed to be held in trust for the community.

The Mier Rural Areas Act also validates past subdivisions of previously communal land into so-called 'economic units', and allows them to be sold privately in the future. In the Surplus Peoples

Project's excellent recent analysis of the history and future prospects of the reserve, a Mier resident complained that the law 'allows private ownership of our inheritance and birthright ... [it] deprives us of our rights' (Wildschut and Steyn, 1990).

The case of Mier bodes ill for the people of the Namaqualand reserves. The boards responsible for some of these reserves have been attempting to use their administrative powers to advance the cause of privatization.

The Labour Party has constantly used the rationale that privatization will curb overgrazing, but this conflicts with the findings of a number of scientists and land use experts. For example, according to Prof Eugene Moll, head of UCT's Botany Department, all the evidence offered in the latest literature on arid lands claims that, through sustainable grazing management, conservation is best carried out under communal ownership.

NAMAQUAN UUS

ABOVE
Tentedorp, outside Port Nolloth, where African people have been living since 1979 when they were separated from coloured people in Paraffienstraat. Tentedorp residents won the legal right to remain in Namaqualand in 1990, but are still waiting for land to be set aside for them

African people in Namaqualand

Falling employment in mining and agriculture has had important consequences for African workers in the region. Under apartheid, the western half of the Cape Province was declared a 'coloured labour preference area'. Despite the scrapping of this policy, no part of Namaqualand was designated for the residential use of Africans. Yet those who lost their places on the mining compounds and farms sought to remain in the region where they had been working for many years.

The result has been squatter settlements in Port Nolloth, the largest coastal town, which is run by a conservative white municipality. The municipal authorities have consistently sought to demolish the settlements and evict the inhabitants from the region. The nearest 'group areas' designated by the state for their use are in Cape Town, 700 kilometres to the south, or Upington, over 500 kilometres to the east. The African residents have resisted these attempts to deport them. Supported by the courts, which granted them permission to settle permanently in May 1990, they have faced obstruction from the municipality, which still refuses to meet residents to discuss the demarcation of a township.

The National Parks Board

An attack from an unexpected quarter came from the National Parks Board, which decided to declare the largest part of the northern Richtersveld reserve a national park. With under 3 000 inhabitants, the reserve is one of the few areas left in the country where some Nama is still spoken. The reserve is rich in minerals and has impressive geological formations, with mountains rising from the desert bordering the canyons carved out by the Orange River as it flows to the Atlantic.

The Parks Board initially insisted that the farmers and pastoralists would have to leave the area demarcated for the national park. The establishment of the park itself was aimed at guaranteeing the active conservation of a fragile local ecology, tremendously rich in

succulents and, in particular, the famous !khureb or halfmensboom, a rare cactus-like plant (*Pachypodium namaquanum*) which grows to approximately 2 metres in size and has a semi-human appearance.

When the community first heard that some of its pastoralists could be deported in the interests of conservation, one man commented: 'Hulle gee om vir die halfmens, maar wat van die volmens?' ('They care about the half person, but what about the whole person?') (Surplus Peoples Project, 1989).

The Parks Board for years dealt exclusively with the local government structures and with the management board of the reserve, and not with the community as a whole. The management board had given its assent to the removals and to the provision of inferior and unfamiliar land for those displaced.

However, the community decided to challenge the contract between the management board and the National Parks Board. On behalf of the Community Committee that was formed to meet this challenge, on 20 March 1989 Mr Willem de Wet applied for an interdict to stop the contract being signed. Within a month the Parks Board and various government bodies began a process of direct negotiations with the community.

Using the same combination of environmental consultants, lawyers from the privatization cases, and the Surplus Peoples Project, the community successfully negotiated the right to remain, to continue grazing its stock in the area, and to receive royalties and jobs. But most important of all, the community persuaded the Parks Board that they should have a large say in the running of the new park and be able to review the agreement in its entirety after a period of thirty years.

Faced with an average income of R90 per month, the community saw its partnership with the Parks Board as a means of ensuring support in protecting local natural resources. As successful guardians of these resources for many years, the community needed to be supported in its endeavours in a way which nevertheless continued to guarantee its nomadic and communal grazing economy.

'Hulle gee om vir die halfmens, maar wat van die volmens?'

Jobs outside the reserve are scarcer, and a new seven-year drought cycle has just begun, according to Fiona Archer, an ethnobotanist who has worked closely with the community. The royalties of R80 000 a year from the Parks Board are destined for essential community development in an area where there is hardly any infrastructure. 'At present, to get petrol,' Archer points out, 'involves a long trip on hard roads to Alexander Bay, 50 kilometres away from the reserve. There is no electricity, but the people are blinded by the bright lights of the nearby mines at night' (personal communication, 1990).

The park will change some of these conditions. The creation of an infrastructure, the royalties, and the jobs offered preferentially to local people, will all contribute to their survival. More importantly, the park offers sustainable development in the area for the first time, instead of jobs in a depleting mining sector. And sustainable development is crucial to the survival of the Richtersvelders. Among the last people to embody the vestiges of an almost extinct Nama culture, their fate is as important as that of the rare flora.

What is particularly significant is the recognition afforded by the Parks Board, which underwent a conversion in its ideology during the negotiations. The Richtersveld National Park is planned as the first Schedule Five park in South Africa, one in which the inhabitants will be integrated into the processes of management and conservation.

It is a tribute to the community that this kind of agreement was negotiated. The ANC commented that 'the joint management of the park by the Parks Board and the community bodes well for a future approach to the environment' (ANC Western Cape, press release, 9 November 1990).

Due to be signed on 10 November 1990, the landmark agreement was sabotaged by the Labour Party when the responsible Minister of Local Government and Agriculture, Rev Andrew Julies, refused to attend the signing ceremony. By virtue of his position, Julies is the trustee of all Namaqualand and other 'coloured' reserves. Without his assent, the agreement is invalid.

Citing the Holy Spirit as his inspiration for not attending the ceremony in the settlement of Khubus, Julies also expressed bitterness because the community had invited Lala Steyn, the Surplus Peoples Project rural coordinator, and a stalwart source of support during the negotiations, to share the ceremonial platform.

The outcry against Julies came from conservationists, scientists, and communities all over the world. The Richtersveld Community Committee attended public meetings in Cape Town and received greetings from other embattled communities, including the Pitjantjatjara people, who form a majority on the board that manages the Uluru-Katatjuta National Park at Ayers Rock in Australia.

'There is no electricity, but the people are blinded by the bright lights of the nearby mines at night'

— Fiona Archer

Commenting on Julies' response, Richard Hill, a geographer coordinating an interdisciplinary research programme on the Park, said: 'The Richtersveld National Park would have placed South Africa at the forefront of new approaches to conservation adopted by the International Union for the Conservation of Nature in their June 1990 update of the World Conservation Strategy, which recognizes that people must be integrated into conservation initiatives' (Hill, statement, 9 November 1990).

The sabotage of the agreement was seen as a slight not only to the people of the northern Richtersveld, but to all concerned with both the environment and the empowerment of rural communities.

Julies later resigned and was replaced by Minister David Curry, who appears to have adopted a more cooperative approach to the reserve. Residents of the area signed a final agreement to establish the park on 20 July 1991.

Nuclear and toxic waste

During 1989 a private company, Peacock Bay Environmental Services, announced plans to construct a toxic waste incinerator at Peacock Bay, 20 kilometres south of Alexander Bay. Namaqualand was chosen as a site for this hazardous installation on the grounds of its low population density. (See Chapter 11.)

The disposal of the toxic waste generated by industrial activities is closely controlled in many countries. Clearly, the onus should be on the producers of the waste to store it securely on site in conformity with the principle that the polluter pays. Ultimately, national environmental policies and international conventions should be working towards the tailing off of toxic waste production.

However, certain entrepreneurs have sidestepped these principles through the exportation and dumping of unsafe waste in developing countries. The toxic waste profiteers have attempted to trade on the ignorance of governments, tempting them with a chance to earn desperately-needed foreign exchange in the short term, without sufficiently emphasizing the high cost of the adverse human and ecological consequences in the long run. Attempts have therefore been made to dump toxic waste in certain African countries, which lack the technical and administrative capacity to manage and dispose of waste in an environmentally sound manner. On discovering this, some African governments have attempted to outlaw such dumping, both overt and clandestine.

When the AEC conducted studies for siting nuclear waste dumps, it plotted circles 50 kilometres in radius around white-ruled municipalities to avoid placing waste near them

South Africa needs to work out its own solutions to the problem of toxic waste disposal, but it does not generate sufficient toxic waste of its own to warrant a local incinerator. Four-fifths of any toxic waste processed at Peacock Bay would have to be imported. The promoters of this incinerator originally stressed the high foreign exchange earning potential of the project; but what they did not stress is the fact that there would be a toxic residue after processing amounting to more than the original amount of toxic waste generated nationally.

Serious questions remain about the safe transport and disposal of the waste, and the ecological consequences of incineration fallout, of which Namaqualand will be the principal target.

Led by Mr Sidney Saunders of Peacock Bay Environmental Services, the consortium of entrepreneurs promoting the construction of the R400 million incinerator admitted to having foreign backing, and claimed that the only people to have been consulted were members of the House of Representatives. It was subsequently revealed that the consortium was chaired by PK van der Byl, a former member of Ian Smith's Rhodesian cabinet, and that the consortium had been set up by a Swiss arms merchant, Arnold Kuenzler, and former National Party MP Peet de Pontes (Koch et

'Must we prepare our own gallows?'

In December 1989 Koeberg Alert, the country's oldest anti-nuclear organization, received a call from the Rev Peter Grove of the NG Sendingskerk in Komaggas, requesting its attendance at a public meeting in the village.

The people of Komaggas were responding to the possibility of a nuclear reactor being built on what the community regards as its ancestral lands, previously alienated by the giant diamond corporation, De Beers.

The community had learnt that land was being purchased by ESKOM along the coast between Port Nolloth and Hondeklipbaai. If a nuclear reactor were to be constructed, 6 000 people would move into the area from outside the Komaggas reserve, doubling the existing population for the duration of the construction period, a minimum of four years.

The community had formed a civic organization in July 1989 and, together with the church-based Standing for the Truth Committee, invited the Koeberg Alert delegation to address its members. Since 80 per cent of the workers in the community are members of the National Union of Mineworkers, the union was also invited to attend the meeting. Press coverage by the *Namaquanuus* regional newspaper was also arranged.

The Koeberg Alert delegation outlined the dangers of nuclear energy posed to the community, and answered questions raised by the audience. ESKOM's promise of electricity was greeted with derision. In the 1970s, ESKOM had built power lines across the Komaggas reserve for the diamond mine at Kleinzee, but the Komaggas community had received no electricity.

The majority of the jobs on the construction site were unlikely to go to members of the community, since ESKOM's practice is to rotate employees from other sites. 'We need the jobs,' someone said, 'but how many of us will they employ, and how many jobs will go to experts?'

The health and safety aspects were also causing concern. One worker said he had worked at Koeberg power station as a security worker, but that 'I was dismissed because they said the radiation level in my body was too high. Now I don't know whether I will ever be able to have children.' Others felt that the health risks to the community and its future generations counteracted the need for jobs. 'Must we prepare our own gallows?' asked Oom Oulak, a community patriarch.

The nuclear issue was not the only one debated. The question of toxic waste disposal in Namaqualand was dealt with by members of Earthlife Africa, who had accompanied the Koeberg Alert delegation.

The community agreed that it had to respond to these challenges and decided to hold a march to draw attention to the question of nuclear power. Tannie Hannah Maarman voiced the feelings of the community when she said: 'We will fight. I'm not afraid to stand up because I stand for the land.'

In June 1990, Koeberg Alert was back in Namaqua-

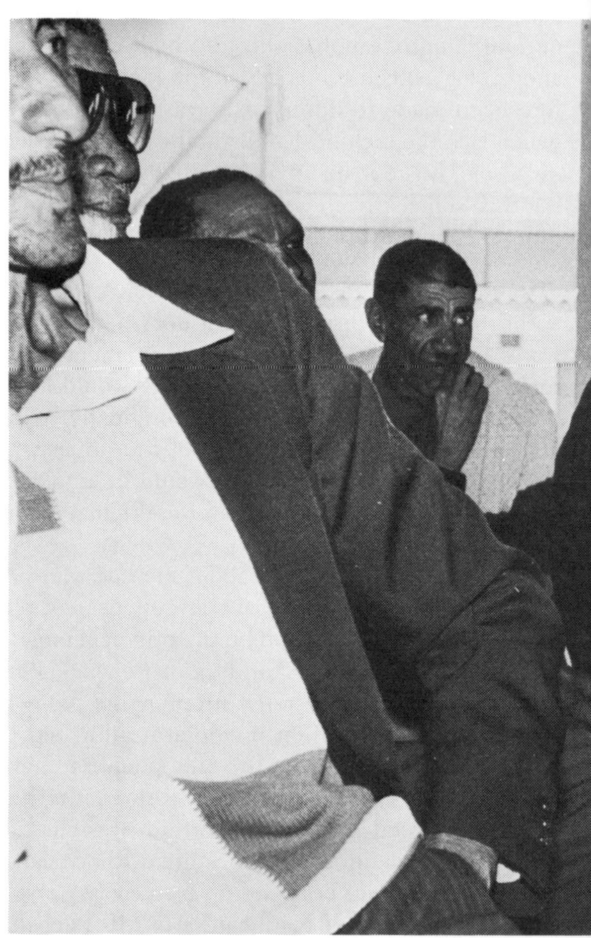

land, this time to address the Nourivier community in the Leliefontein reserve. Nourivier is approximately 28 kilometres from the Vaalputs nuclear waste dump. The people here had become increasingly concerned about the long-term effects of the waste dump upon their lives.

Despite claims by the Atomic Energy Corporation (AEC) that inhabitants were not reliant upon borehole water in the vicinity of Vaalputs, the Nourivier community claimed that both their stock and they themselves relied on borehole water. They expressed disquiet about health and safety practices at Vaalputs, the possibility of having to move the community away from the immediate area, and the lack of any attempt to consult them on the part of the AEC.

The community agreed to get in touch with representatives of the other Leliefontein settlements to draft a memorandum of complaints for submission to the directors of Koeberg.

'Why should we accept that radioactive waste be buried close to our homes when we aren't even supplied with electricity?' asked the community spokesperson, Oom Japie Bekeur.

Within weeks, the matter of nuclear waste had been discussed in the communities of Paulshoek, Rooifontein, Klipfontein and Leliefontein. The memorandum was drafted, and the community planned to circulate it to inhabitants of all the other reserves in the region, hoping that the Namakwalandse Burgervereniging or NBV, the regional civic structure, would take up the problem and begin to lobby for a different solution.

The director of the Vaalputs facility opened up the plant to inspection by *Namaquanuus* reporters and their Koeberg Alert consultants. However, he insisted that any reports be submitted to the AEC in Pelindaba for clearance before publication. When *Namaquanuus* published its report on the visit to Vaalputs it did so without acquiescing to this form of censorship. 'The people have a right to know what we saw, and we have a right to report it,' claimed Elizabeth Beukes of *Namaquanuus*. 'We learnt that there was no consultation whatsoever with the people of Leliefontein when the site was being selected. For the AEC, our people don't seem to exist.'

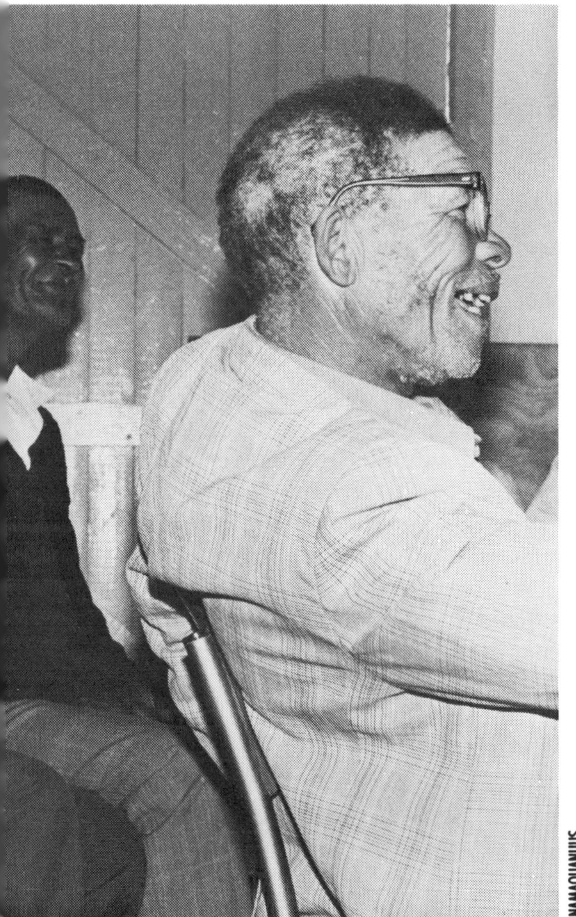
NAMAQUANUUS

Bringing to public attention the nuclear industry's plans for Namaqualand will be a difficult task. Although sites for the next South African reactors have been identified elsewhere, it is more than likely that the Namaqualand site will be the first to become operational. Within the next ten to fifteen years, the people of Namaqualand might be living within the shadow of new reactors. In about forty years the reactors will have reached the end of their operational existence and will be decommissioned, but the land on which they were built will be radioactive for many more years.

Unless the people of Namaqualand act soon, the power of ESKOM and AEC will prevail unchecked. The environment of Namaqualand will be threatened with new dangers. The people of Komaggas and Leliefontein have made a start in challenging the nuclear industry. It is up to all those who care for the land to offer them full support. ❒

For further information contact the Nuclear Working Group, Earthlife Africa, Box 176, Observatory 7935.

LEFT

The Leliefontein committee — Oom Japie Bekeur is on the right — at a meeting in Port Nolloth

ESKOM built power lines across the reserve for the diamond mine, but the community received no electricity

al, 1990). In a separate case, De Pontes was subsequently removed from his seat and jailed for fraud and connections with mafioso mobster Vito Pallazzolo.

The public outcry against the Peacock Bay scheme was substantial. It caused the former Minister of the Environment, Gert Kotze, to announce the investigation of the issue of toxic waste disposal. Kotze asked the parastatal Council for Scientific and Industrial Research to commission a major research enquiry into the question. He announced that during the investigation there would be a suspension of toxic waste imports. This halted any immediate plans to construct the Peacock Bay incinerator.

Despite these measures, it is well known that toxic waste imports have continued. This was highlighted by the case of the Thor Chemicals plant in Natal, which processed imported mercury, traces of which were found downstream in the Umgcweni River. Unconfirmed plans by 'homeland' governments in the Transkei and Ciskei to import toxic waste have also been aired in the press. However, South Africa has now ratified the Basel Convention, an international treaty which regulates closely the transborder shipments of toxic waste. South Africa will become a signatory to the convention once local legislation has been adjusted.

Namaqualand is by no means yet safe from becoming a repository of hazardous waste because the country's nuclear industry has selected the region to house its radioactive waste. The parastatal Atomic Energy Corporation (AEC) in 1987 opened its Radioactive Waste Disposal Facility at Vaalputs, located between Gamoep and Platbakkies, about 100 kilometres from Springbok. Because of widespread international scientific uncertainties about the safe disposal of nuclear waste, Vaalputs remains one of the very few nuclear waste disposal sites in the world. Anti-nuclear activists remain extremely cynical about claims by ESKOM that nuclear waste is 'a problem solved' (ESKOM, 1985).

When the AEC conducted feasibility studies for determining the location of the site, it plotted circles 50 kilometres in radius around the white-ruled municipalities to avoid placing any waste near these areas. However, no similar precautions were taken in the case of settlements in the reserves.

For example, the villages of Paulshoek, Leliefontein, Nourivier, Rooifontein and Kammassies, all in the Leliefontein reserve, are between 25 and 47 kilometres from the Vaalputs site. No-one in the reserves was consulted at any stage about the selection of the site. Even the white-run municipalities were ignored. Eli Louw, the MP for Namaqualand in the (ethnically white) House of Assembly dismissed the issue by reminding critics that 'the people of Cape Town and Blouberg were not asked whether they wanted the Koeberg nuclear power station' (*The Argus*, 19 March 1983). Under the provisions of the 1982 Nuclear Energy Act the minister need not disclose reasons for any nuclear decision on national security grounds.

Louw was referring to the complex 31 kilometres north of the city of Cape Town, whose two reactors have generated sufficient waste for trucks to have to make the long journey to Vaalputs up to four times a week.

Since November 1986, low and intermediate level radioactive waste drums have been transported three to four times a week from Koeberg to Vaalputs for disposal. It was only as a result of concern about a fire which broke out on one of the trucks transporting waste in January 1990, which was voiced in the national and local press, that the people of the Leliefontein reserve began to question the significance of Vaalputs.

Another project earmarked for Namaqualand is the construction of nuclear reactors along the coast between Port Nolloth and Hondeklipbaai. For this purpose, ESKOM intends purchasing land which was originally alienated (possibly under dubious legal circumstances) from the Komaggas Reserve.

ESKOM has consulted very few Namaqualand residents about these plans, which form part of a scheme to build more nuclear reactors along South Africa's coastline over the next few decades. The AEC calculates that South Africa's extensive but finite coal reserves will start running out in the middle of the next century, and that nuclear energy is the only resource that is capable of bridging the gap.

Many environmentalists contest this calculation, and point to the present excess capacity in the electricity industry, the possibilities of alternatives such as solar energy in remote areas, and the advantages of energy conservation. At the World Energy Council held in Harare in November 1990, Zaïre agreed to unleash its huge hydroelectric potential for the use of the entire southern African region. If developed, this will be a source of power equivalent to Africa's entire electricity output at present. It will make the need for nuclear power obsolete in economic terms, although there are some problems attendant upon hydroelectric power in the subcontinent. (See Chapter 7.)

Despite these powerful arguments against the nuclear industry, the anti-nuclear lobby is still extremely weak in South Africa, partly due to legislation governing protest and freedom of information. This enables the AEC and ESKOM to continue with their plans without public accountability.

Having investigated a number of coastal sites, ESKOM has ruled out the Natal coastline, has purchased land in the Oyster Bay-Thyspunt area of the eastern Cape, and has an option on a site at Bantamsklip in the Cape Agulhas area. It is currently investigating sites on the Namaqualand coast, and is apparently focusing its interests on a site 140 kilometres south of Port Nolloth (*Rapport-Tydskrif*, 8 April 1990). Because of the proximity of such a site to the Vaalputs nuclear waste dump, it is likely that the corporation's plans for the Namaqualand site will materialize sooner than those for other areas.

The South African nuclear programme is disproportionately costly and potentially a major threat to human activity and the ecol-

'I was dismissed because they said the radiation level in my body was too high. Now I don't know whether I will ever be able to have children'

— Komaggas worker

The glut in electricity production in South Africa means that the electrification of the reserves could proceed without any necessity for nuclear power stations in the region

ogy; and the link with grandiose military strategies bears investigation. Koeberg is not only home to two economically unviable nuclear reactors, but also houses the training school for kitskonstabels, armed police trainees who perform a vigilante function in South Africa's black townships (Koeberg Alert Research Group, 1987).

The nuclear reactors will have a negative impact on the immediate environment, which is already under siege. They will alienate more land and their construction will lead to the temporary influx of large numbers of people and vast quantities of equipment. It is unlikely that local inhabitants will be employed in the plants after their construction. Nor are the inhabitants of the reserves likely to benefit from the electricity generated, since their settlements are not usually linked to the national grid.

Recent findings by UCT's Energy Research Institute have shown that the electrification of most settlements in the Namaqualand reserves is economically viable in areas currently dependent on more expensive and less convenient fuels. A major obstacle is ESKOM's hesitation. The corporation fears it may not recover the relatively high costs of wiring individual homes, but the Energy Research Institute believes that this could be overcome by using a pre-paid metering system which makes it easier for low-income households to afford being linked to the national grid (Borchers, Archer and Eberhard, 1990).

The glut in electricity production in South Africa means that the electrification of the reserves could proceed without any necessity for nuclear power stations in the region. Given the ideal conditions for generating solar power, there is no need for a significant dependence on the national grid.

The galvanization of the community

Faced with this series of outside assaults, the people of the reserves, townships and informal settlements in Namaqualand have begun to organize themselves.

On the mines, the rapid spread of trade unionism since the late 1970s has made the National Union of Mineworkers (NUM), a powerful voice in the politics of the region. The NUM has attracted the support of some 80 per cent of the region's mineworkers.

According to regional NUM secretary, John Tuck, discriminatory measures are still in force on the mines. Local mineworkers are confined to the lowest job gradings, and there are no members of the Namaqualand community in the senior management of the mines. Miners are forced into a situation of migrant labour. 'There are still a number of fights to face,' he stresses. The dismissal of miners at the Bontekoei mine in the Komaggas reserve in November 1990 is a case in point (Personal communication, 1990).

The setting up of authentic community committees to combat the corruption and despotism of the official management boards has already been noted. In some cases, like the Komaggas reserve, a relatively progressive management board was supported by the community for a time, but was then ostracized and starved of resources by the central government. During 1990 the situation in Komaggas reverted to the pattern of the other reserves.

In informal settlements such as Tentedorp and Bloukamp in Port Nolloth, civic organizations have emerged to defend the rights of African people to remain in these areas.

The civic and community committees throughout the region have set up a regional representative structure known as the Namakwalandse Burgervereeniging (NBV), or Namaqualand Civic Association. The establishment of the NBV was a major step towards the integration of regional struggles and campaigns. Festivities and speeches accompanied the launch of the NBV in July 1989, and the speakers included UDF and SWAPO leaders.

The social role of the churches has been enhanced by the recent

BELOW
In 1990 7 000 people led by local community and church leaders streamed through the streets of Springbok in a May Day march — the first in Namaqualand

WELMA ODENDAAL

establishment of the Namaqualand Council of Churches (NCC), with its headquarters in Springbok. The NCC has been able to utilize its meagre resources to facilitate the mobilization occurring within the communities of the region.

For example, the NCC is partly funding a monthly regional newspaper, *Namaquanuus*, which is brought out by a small group of young trainee journalists drawn from the community and the Cape Town-based Media Training and Development Fund. This paper has played an important part in highlighting the struggles of the different communities, and in establishing links and helping to counteract the isolation the communities face. *Namaquanuus* is currently battling to survive financially, and its demise would be most unfortunate in terms of the progress of the region.

Since its unbanning on 2 February 1990, the ANC has managed to establish a presence in the region, and organized a successful and unprecedented march of 7 000 people through Springbok. Namaqualand forms a sub-region of the Western Cape ANC, and because of its special conditions, has merited the appointment of a sub-regional organizer.

The volume of work in Namaqualand encountered by the Surplus Peoples Project in terms of research, paralegal and community development work, has necessitated the organization setting up a branch office in the region.

Environmentalists and environmental organizations have also been a part of the process of political galvanization in the area. Botanists, social anthropologists, nuclear experts, energy researchers, land use specialists and environmental lawyers have all been approached by communities seeking assistance in their struggles. A key problem is that the sources of expertise mostly lie outside the region, usually in Cape Town, which is between five and nine hours away by road.

Despite these imbalances, the enthusiasm of the trade unions and the communities in defending themselves from the various assaults upon them, has gained them enormous support. With little political experience they have begun to challenge the mining companies, corrupt management boards, the privatizing cabinet ministers, the white-run municipalities, and bodies like the National Parks Board, the AEC and ESKOM.

With limited resources available, the Namaqualand communities have succeeded in some of their objectives. Despite their unique character, their struggles are a microcosm of the struggles of all rural South African communities, marginalized because of their isolation, distance from urban centres, and their small numbers. Their demands for land, for basic services, for communal grazing and for the right to remain in the area in peace, are matched by their rejection of dummy local authorities, of privatization, of the dumping of nuclear and toxic waste, of the state's nuclear and military expansion in the region, and of exploitation on the farms and mines.

As communities assert these demands, they become aware of the need for unity, for extra-regional networking, and for the gaining of political experience to assist their endeavours. The benefits of all these strategies have been tangible.

It is the land question that is central to solving the problems of Namaqualand. Unless the need for its equitable distribution, conservation, and sustainable development is taken seriously, creeping aridity and depleted resources may undermine all possibilities for social renewal.

When the spring flowers of Namaqualand are next mentioned, think of them as an analogy of the struggles there. Amid the harshness, the dryness and meagre resources, a sudden flowering that transforms the landscape is possible. And like the fragile ecology, these struggles will need committed support.

9

WASTING WATER

Squandering a precious resource

◆ HENK COETZEE ◆ DAVID COOPER ◆

South Africans pride themselves on a land endowed with great national riches. It provides food and minerals to many countries, African and other, and is famous the world over for its natural beauty. However, one resource, possibly the most important of all, is severely lacking: water. Large areas of South Africa are semi-arid or arid, receiving a rainfall of less than 250 millimetres per annum.

BELOW
Collecting water in Driefontein in the eastern Transvaal

GILL DE VLIEG

What little water we have faces the grave threat of becoming unusable as industrialization proceeds virtually unchecked

The mountainous escarpment areas have a high rate of run-off, and ensuing problems with soil erosion. What little water there is faces the grave threat of becoming unusable as industrialization proceeds virtually unchecked.

Water is supplied to its points of use largely from dams, as South Africa has few natural lakes or large rivers. The country has at present a total of 519 dams with a total capacity of 50 000 million cubic metres. However, the arid climate, frequent flooding and poor management have resulted in the silting up of many of these dams.

Water from these dams is pumped to the industrial areas, many of which are situated far from their water sources. Water is also pumped from underutilized river systems to other systems where water is in short supply. The Highlands scheme, for example, will eventually divert over 2 000 million cubic metres of water from the headwaters of the Orange River into the Vaal River system. This and other schemes will ultimately provide 75 per cent of the Vaal River's water.

South Africa does not only face severe water shortages; the quality of the water also poses problems. South Africa, in common with other industrializing Third World countries, is trying desperately to develop in accordance with First World patterns, but uses legislative and administrative controls over industry that follow Third World patterns. While there are modern chemical plants producing toxic waste, the state does not have the laws, staff or infrastructure to prevent them from despoiling the natural environment. The social infrastructure is also not well attuned to opposing environmental destruction; the black majority has a history of concern and involvement with more apparently political struggles, whereas the white minority has largely remained sufficiently complacent to allow the preservation of the status quo.

Water: where, what and why?

Almost all water for consumption must be drawn from rivers or underground reservoirs. For purposes of large-scale water supply, water must be stored in dams. South Africa is also subject to droughts and a reliance on rainwater alone even for agriculture is extremely risky. Rivers and lakes are the most heavily utilized sources of fresh water and must therefore be protected from degradation, both of their ability to keep up with demand and of the quality of water that they carry. Underground water too must be seen as a limited resource, and the risk of contamination should be taken into account when new developments are planned. (For a discussion of the effects of strip mining for coal on underground water, see Chapter 7.)

Water enters the system as rainwater, either percolating into the ground and entering underground reservoirs, running off into river systems, or evaporating. Much of South Africa receives rain in the form of summer thunderstorms, and this water tends to enter rivers rather than underground water systems. Unless this water is utilized it will be lost into the sea.

The Vaal River system, comprising a network of dams, rivers, streams and canals, is probably the river system most important to the economy of South Africa. Carrying only 4 500 million cubic metres of water per annum, it is actually a small river system compared with, for example, the Zambezi at Cahora Bassa, which carries 88 000 cubic metres. Yet the Vaal River system supplies water to the whole of the Pretoria-Witwatersrand-Vaal (PWV) area: 40 per cent of the population of South Africa and 60 per cent of industrial production thus depend on it. It supplies the goldfields of the West Rand around Klerksdorp, the Orange Free State goldfields, the Vaal-Harts irrigation scheme, Kimberley and dozens of small towns in the Transvaal. It also carries much of the liquid effluent that these industries and towns produce.

South Africans experience a sense of disempowerment regarding most environmental issues

The degradation of water resources

Water resources become degraded in two main ways. The quality of water can be contaminated by pollution, and the supply of water can be seriously affected by the abuses of the environment which take place in the river catchment areas and along the rivers themselves.

Large areas of the eastern Transvaal Lowveld were once fertile valleys where people lived and farmed productively. With the establishment of commercial timber plantations in the catchment areas of the streams and rivers which feed these areas, the rivers have dried up and the soil has been leached of moisture. Regions which used to support their populations are now dormitory areas for the families of migrant workers who must travel to the cities in search of work.

Degradation of the rural areas owing to overcrowding of the government-created 'homelands' is seen in soil erosion and deforestation. The degraded soil is washed away by the rainwater instead of

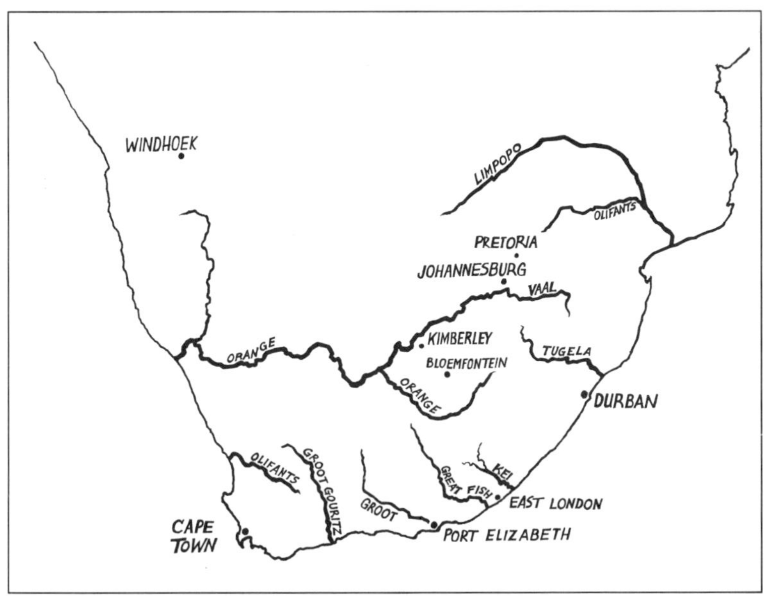

LEFT
South Africa's main rivers

South Africa is trying desperately to develop in accordance with First World patterns, but uses legislative and administrative controls over industry that follow Third World patterns

being replenished by it. Water and soil are thus lost into the sea. (See Chapter 12.)

Siltation from soil erosion is a serious problem throughout South Africa. Annually, it causes the loss of 130 million cubic metres of storage capacity, nearly equal to the loss of one medium-sized dam such as Midmar or Hartebeespoort. Siltation itself is a natural process, whereby soil is washed from the land into its rivers. But soil erosion arising from bad farming practices leads to excessive siltation, thus affecting biological processes in rivers. Soil accumulates in dams, decreasing storage capacity and also raising the rate of evaporation.

There are a number of ways in which water can be polluted. Acid rain, fertilizers, pesticides and municipal dumps all contaminate our surface water resources.

Highly acidic rainwater frequently falls in the industrial heartland of the eastern Transvaal Highveld. This pollution comes from the power stations and industries which rank this area with the most polluted in the world. (See Chapters 7 and 11.) Although acid rain is attacking forests, the quantity of pollutant reaching the rivers is not yet significant, because of low run-off and the wide area of dispersal.

Farmers improve crop yields with a range of agricultural chemicals, but these also wash out of the soil and into river systems. The nitrates and phosphates in fertilizers promote plant growth in rivers. These plants use up the oxygen in the water, starving out other aquatic life. This process, called eutrophication, was responsible for the uncontrollable growth of water hyacinth in the Hartbeespoort dam in recent years.

Pesticides, insecticides and herbicides are poisons. Residues are washed into underground water systems and rivers and can even be carried in rainwater. High concentrations of the hormonal herbicide 2,4-D have been found in rivers in Natal, adjacent to the sugar cane fields where it is used. Farmers in the Tala valley in Natal have repeatedly suffered severe crop damage from a mixture of the hormone herbicides 2,4-D and 2,4,5-T, the same ingredients used in the infamous Agent Orange, a defoliant used by US forces during the Vietnam war.

Most modern pesticides have a fairly short lifespan once released into the environment and their use can be reasonably strictly controlled. Older, more persistent pesticides like Dieldrin, DDT or BHC (Lindane) create far more serious problems. They are transferred in the food chain from insects to fish to birds and to mammals, accumulating in fatty tissue. They are also transferred to humans via the food crops that they have been used on. When pesticides reach rivers, they affect aquatic life and users downstream. Traces of pesticide have already been found in the Vaal River system, although concentrations are said to be too low to seriously affect humans.

There is, however, an alternative. Integrated pest management is a system that aims at minimizing the use of pesticides by using natural, environmentally sensitive methods of pest control, such as intercropping and the use of natural predators on pests.

Most developed countries have strict controls over the use of pesticides, and many of the pesticides used freely in South Africa are banned in these countries. Pesticides such as Lindane and 2,4,5-T are banned or severely restricted in other countries and should also be tightly controlled here. Yet some of their most liberal users have been the Departments of Agriculture and Forestry, the agencies which should take responsibility for limiting their use.

THE ARGUS

ABOVE

Dead fish in the Liesbeeck River, which runs through the heart of Cape Town's southern suburbs

Another process affecting surface water seriously is salination. In natural systems, the salts left behind by evaporating water are diluted and washed away by rain. Irrigation, however, does not allow water to wash away, and the salts therefore become increasingly concentrated. A million hectares of arable land worldwide are lost to salinisation annually. Irrigated land on the Transvaal Highveld is affected by salination, and the mining operations in the area are exacerbating the problem by pumping large amounts of salts into the river systems: 60 per cent of the salt load entering the Vaal Barrage is caused by the effluent from four mines.

Municipal refuse dumps and landfills are among the most serious sources of water pollution. (See Chapter 11.) Pollutants can leach from these sites into underground water and surface water supplies. Toxic industrial waste is often merely diluted with domestic waste, much of it also highly toxic, and dumped into clay-lined pits. These are frequently sprayed with pesticides, resulting in a lethal cocktail of toxins. Although these sites are monitored, this monitoring only identifies problems or disasters. It cannot prevent or remedy them.

Communities are growing in their awareness of this type of problem, with civic organizations opposing the development of new landfill sites. One such group is the West Rand Environmental Awareness Committee. Based in Azaadville, this committee is concerned about the creation of a toxic waste landfill in their area, a zone prone to earth tremors and close to dolomite reservoirs which are a potentially valuable water source for the PWV.

Illegal dumping creates more problems, as it cannot be controlled by competent authorities. The Johannesburg Branch of Earthlife Africa is investigating the illegal dumping of toxic wastes on the Witwatersrand and has been alerted to a number of illegal dumps.

Surface water pollution

An investigation by *The Weekly Mail* in July 1990 found that a stream flowing between old mine dumps and a new slimes dam in the Crown Mines area, and then through crowded districts of Soweto, contained exceptionally high levels of pollution. Laboratory analysis of water and sediment samples from the stream showed that the dumps were leaching a staggering 8 million micrograms of sulphate, 1 900 micrograms of uranium, 520 micrograms of cyanide and 60 micrograms of arsenic per litre into the water. These levels far exceed the World Health Organization's standard for 'wholesome water', a standard which is realistically lenient in order to allow development in Third World countries.

Further downstream, the same river was found to contain unacceptably high levels of petrol, leaked from a nearby service station. Homeless people in informal settlements on the river rely on this as a source of water. Local residents, while concerned about the petrol and other river pollution, often do not know how to go about reporting these problems. This is typical of the sense of disempowerment experienced by South Africans regarding most environmental issues. The problem was eventually referred to the Rand Water Board, who undertook the necessary testing and initiated corrective action. Had this case not been reported, the Rand Water Board, which, like so many other state departments is drastically understaffed, would probably never have known about it. All over South Africa, similar pollution is continuing unchecked, owing to an unwillingness to act on the part of the authorities, a sense of powerlessness among the people, distrust of officials and the minimal concern which central government gives to environmental issues when it comes to budget allocations.

Industry in South Africa has much to answer for. In September 1989 the Sappi paper mill at Ngodwana in the eastern Transvaal Lowveld leaked enough sulphate and soap skimmings into the Elands River to kill 100 tons of fish (twenty-two species in all) and most other forms of life in the river. For this the company was eventually fined only R6 000.

BELOW

The SAPPI paper mill in the eastern Transvaal Lowveld

JUSTIN SHOLK

Regular tests conducted by the Rand Water Board at more than forty sampling points in the Vaal catchment area reveal that mine dumps around the SA Land and Exploration Company on the East Rand are leaching traces of mercury, large amounts of arsenic and alarming levels of sulphates, that exceed European standards for surface water, into the Rietspruit stream.

The AECI Midlands plant outside Sasolburg has a government permit to emit effluent with a mercury content of 20 micrograms per litre into a stream which feeds the Vaal River. The London-based environmental group, Friends of the Earth, say that this is more than six times the limit that is legally enforced in European Community countries. What is more, senior AECI officials admit that even this already unacceptably high level is often exceeded. Recent investigations by Earthlife Africa and Greenpeace have also found that water in a stream close to the factory is so polluted that it has corroded a concrete culvert. Inside the plant, workers report that mercury poisoning cases have occurred and that the standards set for mercury poisoning (300 micrograms per litre in urine) are six times those set by the National Centre for Occupational Health.

Also in the Sasolburg area is the giant Sasol I complex, which pumps large amounts of organic pollution into the Vaal River. Tests have identified twenty-seven chemicals listed as priority pollutants in the United States in Sasol's effluent. While the authorities and the industries in the Sasolburg area assure the public that these effluents only enter the Vaal downstream from the Rand Water Board's inlet pipe, this does not help all those downstream who depend on the water. Parys, for example, has levels of trihalomethanes (potential carcinogens formed by the reaction of organic chemicals and chlorine) up to seven times the European Community guide level.

Outside the industrialized PWV area, serious water pollution problems also occur. The scenic Barberton mountainland in the eastern Transvaal is famous for its disused gold mines and historic mining

Acid rain makes the eastern Transvaal Highveld one of the most polluted areas in the world

stories. Unfortunately, there is more to this area than the romance of the nineteenth century gold rush. Chemical engineers describe the mines as arsenic factories, and a dump left by a prospecting team from Anglo-American contained so much arsenic and cadmium that it killed all the aquatic life in a local dam. This dam is situated on a farm that was awarded Natural Heritage Site status.

In general, pollution in South Africa is dealt with using the 'best practicable means' (BPM) approach. Here industry decides what means it can apply to stop pollution, and the relevant government departments then apply standards which reflect these methods. Industry thus sets the agenda for pollution control. Permits are issued which allow companies to emit higher than legal concentrations of pollutants. The system of measuring emission concentrations is also flawed. Most developed countries measure total pollutant quantities emitted, rather than concentrations, as these allow companies to simply dilute their wastes down to an 'acceptable' concentration.

The problem is not insoluble. The difference between the older Sasol I plant, and the newer Sasol II and III plants, is striking. Sasol I discharges toxic waste, while II and III discharge no waste water at all. Improved plant design means that chemicals are recovered and clean water is recycled to the plant. Although these options are not open to all plants, especially small-scale ones, it is a feasible objective that all effluent waste leaving factories should be non-toxic. This clean technology can also benefit the industry, as the recovered material can be a valuable asset.

The water that we drink

The government and industries want to see South Africa become a 'First World country'. At the same time they hold one of the basic determinants of quality of life in contempt. Clean water is a necessity for survival. However, in South Africa it is a privilege enjoyed by an ever-decreasing number of people. The standards set for the quality of drinking water and its distribution inside South Africa do not tally with First World conditions.

In June 1990 *The Daily Mail* carried an exposé of mercury levels in Johannesburg tap water — they exceeded the WHO and European Community standards

It has been said that the health of a community is best measured by the number of standpipes rather than by the number of doctors, and since as much as 80 per cent of all disease under Third World conditions is related to poor and inadequate water supply, this is reasonably accurate. The Winterveld squatter camp outside Pretoria, for example, has no installed water supply for its population of over half a million. Water is sold at prices of up to twenty cents per bucket, and the lack of any sanitation means that water boreholes are frequently contaminated with human, animal and domestic waste.

Not only is the quality of South Africa's water erratic and often poor, but its distribution is also unequal. The following figures indicate the amount of water used per person per day in various parts of South Africa, expressed in litres.

AREA	PER CAPITA CONSUMPTION
Ciskei	9
Eastern Cape townships	19
Port Elizabeth townships	80
Average white suburb	200-350
WHO standard	50

Any new South African government will have to take this information into account when looking at redistribution of resources.

In white urban areas, a disaster may be brewing. The poor standards, lack of their enforcement and the contempt shown by industry for what standards do exist, place at risk the ability of the water authorities to provide a clean drinking water supply. In June 1990, *The Daily Mail* carried an exposé of mercury levels in Johannesburg tap water, which exceeded the WHO and European Community standards for drinking water. Bacterial contamination of water is neutralized with chlorine, but this process allows the formation of the carcinogenic trihalomethanes and chlorobenzenes.

Trihalomethane levels in Johannesburg tap water are generally within the South African Bureau of Standards (SABS) guideline of 100 micrograms per litre, but elsewhere the situation is more serious. Levels of up to 765 micrograms per litre have been measured in Parys, and a recent Council for Scientific and Industrial Research (CSIR) report notes that tap water in Hammanskraal and Pienaars River contains levels which are consistently higher than 200 micrograms. The European Community guideline for trihalomethanes is 1 microgram per litre, and many experts now believe that there is no safe limit for these chemicals. There are alternative processes to the chlorination process, such as ozonation, but water authorities in South Africa do not acknowledge that there is any cause for concern and are therefore not

BELOW

Part of Earthlife Africa's protest against the quality of Johannesburg's drinking water in June 1990

AVIGAIL UZI

Unlike other industrial countries, South Africa has no legislated standard for drinking water

considering applying safer water purification techniques.

Unlike other industrial countries, South Africa has no legislated standard for drinking water and, contrary to normal practice elsewhere, the guidelines have been relaxed over the past few decades.

A recent report by Paul Polasek and Claude Mangeot, two engineers concerned with the quality of drinking water, notes that the SABS 241/84 specification for drinking water does not set standards for a number of potentially hazardous contaminants in drinking water. Standards which are set are often more lax than those in other countries, and since SABS 241/84 is a guideline and not a standard, it is unenforceable. These inappropriate specifications have led to a lax attitude on the part of the water purification authorities, and resistance to new technologies such as the ozonation technique discussed above. The SABS 241/84 specification is a more lenient version of an earlier specification, which appears to have been deliberately relaxed in order to accommodate the increasing levels of pollution by the chemical industry.

What of the future?

Apartheid has clearly affected the availability of clean water to the majority of South Africans. People in rural areas and urban ghettos do not have adequate water supplies, and what water there is is often contaminated with industrial and domestic pollutants. This has made water supply an area which is often hotly contested within communities; it has become a focus for many community struggles.

It is often suggested that the transition to a democratic system will bring water to the people, but a more practical approach than this is needed. People must take responsibility for ensuring that they have an adequate supply of clean water, and a democratic society must provide the facilities and means for them to do so.There must also be a move away from the current system of land dispensation, which locates marganilized people in areas with insufficient land and water resources. Information regarding water quality, pollutants entering the water and polluters who despoil the water is essential for people trying to ensure a clean water supply. Whether the new political dispensation will provide people with this information is a critical issue.

Another essential task of any new government must be the fuller participation of the affected communities in new water supply schemes and new industries which may pollute. By involving communities at all levels, from planning to implementation of these schemes, a more responsible approach to development can be achieved.

South Africa's water resources are scarce and a shortage inevitable. By the middle of the next century, shortages will be felt by everybody, and even in this decade we are likely to see industrial development being hindered by the shortage of water. Under these circumstances it is extremely short-sighted to allow this scarce resource to become further polluted and degraded. What is essential now is the education of all South Africans in the value and wise use of water, and a government which does not allow the most vital of all of the country's resources to be abused and become ultimately unusable.

10

THE INSANE EXPERIMENT

Tampering with the atmosphere

◆ JAMES CLARKE ◆

The air we breathe

In June 1990 the satellite Voyager I sent back an image of planet Earth taken from 6 000 million kilometres away. There we were — a tiny bright blue speck in the eternal silence, encased in a shimmering but diaphanous membrane of life-supporting air and water. We looked very lonely. As Carl Sagan said, if there is indeed intelligent life out there in the cosmos, then that is an awesome thought. On the other hand, if there is no-one else out there, if we are truly alone, then that is an even more awesome thought. The only truth we have to go by, for the foreseeable future, is that there is only one Earth.

Yet we have allowed this planet's life-support system — its finely-tuned arrangement of air, water and soil — to become seriously damaged. Governments the world over have allowed industry and other interests to conduct a massive and unchecked experiment on the chemistry of Earth's atmosphere, without the faintest idea of what the consequences might be.

The Earth's atmosphere comprises 78,09 per cent nitrogen and 20,94 per cent oxygen. The less than one per cent that is left comprises several gases, the most important of which is carbon dioxide, amounting to a mere 0,03 per cent. It is, in a sense, the tiny linchpin of the whole living system. Plants absorb carbon dioxide in order to convert sunlight (by a process known as photosynthesis) into food. In so doing, plants give off a waste gas — oxygen. Mammals, such as human beings, cannot live without oxygen. In using oxygen we, in turn, give off a waste gas — carbon dioxide — which plants absorb. It all adds up to a very agreeable arrangement.

We are tampering with the makeup of the atmosphere on a radical scale

It should be an article of survival that humankind should never interfere with the makeup of the atmosphere. Yet we are doing so on a radical scale.

Governments the world over have allowed industry and other interests to conduct a massive and unchecked experiment on the chemistry of Earth's atmosphere

Statues, historic architecture and other works of art are suffering damage throughout the world. Venice's elegant cityscape is rotting from air pollution; the Parthenon, above Athens, built in the fifth century BC, is rapidly decaying. The 4 500 year old pyramids in Egypt are crumbling and so is the Sphinx, whose great head is in danger of falling off. The Taj Mahal in India, and the Statue of Liberty in New York, are likewise victims of a steady corrosive fall-out. The picture emerging from eastern Europe is even more horrific: statues and buildings of great cultural value have rotted in the acid rain. In Poland, some look as if they have been burned by fire. Throughout Europe, a third to one half of natural forests are now damaged and large forest areas in Czechoslovakia, Germany and Switzerland are dead.

South Africa has not escaped. An example of how quickly acid rain damage can occur can be seen at the University of the Witwatersrand: Iron Age artefacts which had lain intact in the veld for centuries before being consigned to the university's archaeological collection are now corroding rapidly in the city air. City buildings have to be washed down every five or six years nowadays — each act of cleaning removing another layer of decayed stonework and concrete. Roof guttering firms are doing a roaring trade because gutters last only half as long as they used to. The first signs of acid rain damage are now appearing in the eastern Transvaal, where pine needles are changing from a healthy dark green to a sickly mottled beige.

Yet we are being told by industrialists, health officials and politicians that there is nothing to worry about. They assure us that nobody has thus far found irrefutable evidence that our chemical-laden air is unhealthy. The evidence does not exist for one very good reason: nobody has seriously sought it. Even in Soweto, where researchers found that the air was two and a half times worse than in the notoriously polluted eastern Transvaal Highveld, no attempts to gauge health effects were made until June 1990, when a ten-year-long Birth to Ten programme to monitor the development of children in the Johannesburg-Soweto region was started by the South African Medical Research Council (MRC). Dr Yasmin von Schirnding of the MRC told *The Star* (26 August 1990) that next to gastroenteritis, respiratory disease was the major killer of children under five. She added: 'We also found that acute respiratory problems in children may possibly lead to chronic lung ailments in adulthood.'

There have been only two attempts to assess the impact of air pollution on public health in South Africa. Both were rudimentary, both were alarming, and both were underplayed by all concerned.

The two surveys revealed that boys born in polluted Highveld towns such as the coal-mining town of Witbank in the Transvaal and the petrochemical centre of Sasolburg in the Orange Free State are generally 2 centimetres shorter than boys in clean rural towns. (Female children appear not to be affected in the same way.) There was also evidence that their breathing was affected by air pollution.

In 1984 one of the researchers, Prof AM Coetzee of Pretoria University, compared the physical development and lung functions of 335 Standard Three children in Sasolburg with those of 147 children in the relatively unpolluted towns of Heilbron, Frankfort and Parys. He found that boys in Sasolburg were 2 centimetres shorter than average and showed impaired lung functions.

Coetzee's survey, apparently conducted on a shoestring budget, was too skimpy to be of much scientific value, but it should immediately have prompted the government to conduct a more comprehensive survey. Yet it was not until 1988 that this happened. This time, using a sample of 2 000 children, Prof Saul Zwi of the University of the Witwatersrand compared the development and lung functions of children aged six to ten in industrial towns such as Witbank, Springs and Welkom, with those of children living in 'clean' towns such as Nylstroom in the northern Transvaal and Wolmaransstad in the western Transvaal. Prof Zwi, under the auspices of the Council for Scientific and Industrial Research (CSIR), found that boys in the polluted towns were appreciably shorter than those in country towns. Although their lung functions did not appear to be affected, they were far more prone to early morning coughing, wheezing and asthma than were the children in the rural towns.

There is no 'cancer atlas' to show where in South Africa different kinds of cancer may be clustering, or where respiratory complaints are the most common. In the United States and Europe, even where pollution appears to be less heavy than in South Africa, medical research and epidemiological studies have come up with worrying but vital information. A cancer atlas in the United States not only pinpointed areas where the health authorities had expected cancers to be clustered — Los Angeles, for example, where, because of photochemical smogs, lung cancer is more common than in other places — but it usefully revealed one lonely spot high up in the scenic north. An investigation showed that a copper mine,

Boys born in industry-polluted Highveld and Free State towns are on average two centimetres shorter than their rural peers

JUSTIN SMOLK

unable to sell its arsenate of lead, a by-product of the mine's activities, had been sending it up its smokestack and pumping it into the atmosphere. The death rate from cancer on the plant and in the mining community was found to be way above the national average and the plant was closed immediately. The company had been dumping the arsenate of lead ever since Canada, which had once bought it, had declared it too carcinogenic to import. The episode demonstrated how, even in this apparently enlightened age, industrialists sometimes reveal a chilling callousness.

In Britain a similar study revealed a cluster of leukaemia cases around Sellafield nuclear power station in north-western England. At first a power station leak was suspected; then radon, a gas which exudes from granite in the area, was suspected. Now the theory is that nuclear workers, contaminated by radiation at work, may have induced leukaemia in their children at conception. Whatever the truth may be at Sellafield, this case study does reveal how a national cancer atlas may a good indicator of pollution trouble spots.

In 1989, at Phalaborwa in the north-eastern Transvaal, farmer Antoon Lombard lost forty-five head of cattle from copper poisoning after they had eaten grass heavily contaminated with copper. Their livers contained three to nine times the lethal dose. It was later ascertained that the local mining company's pollution control equipment had been out of commission for some months. The Industrial Hygiene Division of the Department of National Health defended the plant, saying: 'This [contamination] is to be expected in the vicinity of such a large mining complex.' The government official went on to remind the reporter who recorded the statement that the town of Phalaborwa depended on the mine for its economic existence. Time and again one comes across this attitude, that a town's economic needs should override health considerations.

South Africa's reluctance to delve too deeply into the health effects of air pollution is not unique. The British science journal, *New Scientist*, in an editorial in 1989, suggested that people were like frogs: put a frog in a pot of hot water and it will leap out. Put it in a pot of cold water, and bring it slowly to the boil, and it will sit there until it boils to death.

Sulphur dioxide, the killer ingredient in London's infamous smogs and identified by most experts as the major cause of acid rain, is released mainly from burning coal. It converts to sulphur trioxide, which is a powerful acid. The water droplets in London's smogs probably had a pH factor of 1,7 — in other words, they would be extremely acidic. In Cape Town, in the smoggy days before Koeberg nuclear power plant led to the closing of the city's coal- and oil-fired power stations, atmospheric acidity sometimes caused women's stockings to dissolve on their legs.

According to Prof Ian Webster, who addressed the International Conference on Air Pollution of the CSIR in 1979, the size of airborne particles is critical in allowing sulphur to enter the lungs. Most airborne particles are large enough to be trapped in the nasal

LEFT
Early-morning air pollution in Soweto

hairs or expelled by the coughing reflex, but sulphur dioxide is soluble in nasal and throat fluids. In London's mustard-coloured smogs, these tinier particles slipped through the lung's defences and allowed sulphur to come into contact with the lung tissues, upon which the sulphur dioxide acted like battery acid.

In January 1989 Prof Peter Tyson, climatologist at the University of the Witwatersrand, together with other scientists, produced for the CSIR a report which revealed the extent of industrial air pollution over the eastern Transvaal Highveld. The report commented: 'The emission densities are between five and just under ten times greater than those found in West Germany and the United States, and approximate the worst conditions found anywhere.' (The authors included comparative figures from the USA, Canada, East and West Germany, and Britain.)

The report painted a sorry picture. These South African sulphur dioxide emissions were 31,25 tons per square kilometre per year. East Germany, notoriously dirty because of the quality of its coal and the way this coal is burnt, emits 30 tons. One of the United States's largest industrial regions, Ohio, emits 19,8 tons per square kilometre per year. Britain emits 14,3 and West Germany, where years of sulphur pollution and the resultant acid rain are now killing forests and corroding statues, emits only a quarter of the eastern Transvaal volume.

The Tyson report, by its authors' own admission, is cautious. At the time that it was made public a second report, privately commissioned by manufacturers of electrostatic precipitators and other clean air equipment, was published by *The Star*. It held that the eastern Transvaal Highveld sulphur emission density was not 31,25 tons, but 57,5 tons — almost twice as high as East Germany's. A subsequent CSIR figure suggested it could even be 60 to 80 tons per square kilometre. The Department of National Health pointed out that it was unfair to compare the local region (30 000 square kilometres) with a whole nation such as East Germany (108 000 square kilometres) or even with Ohio, which is about the same size as East Germany. In fact, taking South Africa as a whole (all 1,2 million square kilometres), our average national sulphur emissions, according to Tyson's report, amount to a mere 0,8 tons per square kilometre per year. It is true that most of South Africa, most of the year, has beautifully clear blue skies, but it is equally true that sulphur emissions are concentrated in the region that is the most heavily populated — the southern half of the Transvaal.

In the triangle formed by the giant power stations in the eastern Transvaal Highveld, sulphur emissions rival those of the Ruhr Valley

According to Gerard Held of the CSIR, in the triangle formed by the giant power stations at Matla, Duvha and Arnot, just south of Middelburg, sulphur emissions rival those of the Ruhr Valley — around 860 tons per square kilometre per year. Even this last figure was still way below some of the readings coming from Poland and Czechoslovakia, where conditions in 1990 were so smoggy that cars had to switch their lights on at midday. People were evacuated from villages in Silesia, Poland, where cancer was up 50 per cent on the Polish average and lung complaints were up 30 per cent. The trouble with South Africa is that industrialists, as well as the Department of National Health, tend to take comfort from such comparisons. They do not consider the situation in South Africa bad enough to warrant serious action.

Prof Harold Annegarn, chief research officer at the Schonland Research Centre for Nuclear Sciences at the University of the Witwatersrand, is dismissive of the eastern Transvaal Highveld figures. He argues that sulphur dioxide pollution can be divided into two categories: high-level emissions which are discharged through tall stacks and which are meant to disperse harmlessly before they fall back to earth; and low-level emissions, which arise from burning coal dumps, from the smaller power stations with low stacks, from small industries and from black residential areas, which mostly have to burn coal because they have no access to electricity. Annegarn, who is also an office bearer in the National Association for Clean Air (NACA), insists that the only worthwhile statistics are those relating to ground level sulphur quantities.

The country's chief air pollution control officer is not obliged to produce an annual report

But what goes up must come down, somewhere. One does not have to be a scientist to know that pumping one million tons of sulphur a day into the air above a highly-populated region, and into an atmosphere which is stagnant for the four months of winter is, to say the least, a non-therapeutic practice.

The industrialized Transvaal Highveld region is 1 600 to 2 000 metres above sea level, which means that there is a fifth less oxygen available than at the coast. This reduces the efficiency of all combustion in the area. Thus, for example, a diesel vehicle which runs cleanly in Durban may emit clouds of black exhaust smoke once it climbs onto the Highveld.

The Highveld has a second handicap: its temperature inversions. These are particularly pronounced in winter when warm layers of air trap cold below them. A typical winter's dawn sees a layer of cold air trapped at ground level under a roof of warm air. In industrial areas the rising of the sun reveals a seemingly solid brown blanket of pollution.

On the one hand, South Africa is fortunate in having an abundance of coal — enough to last 300 years at current rates of extraction. On the other hand, we are unfortunate in that these coal supplies are located in the Highveld, a climatic zone unsuitable for coal combustion. Giant power stations, fed by coal mines, dominate the eastern Transvaal Highveld's undulating savannah. The coal, mostly of fairly poor quality, is crushed to a fine powder before being poured into the furnaces to make steam. Considering the complicated and useful chemical composition of coal, the day will surely come when such a crude use of coal will be viewed as sheer extravagance.

In 1990 ESKOM commissioned a 3 400 megawatt coal-fired power station — Lethabo — in the Vaal Triangle. Lethabo burns 15 million tons of coal per year. This power station, just to give some idea of its output, could supply all the annual electricity needs of the state of Israel, even with one of its turbines switched off. The Transvaal has eight of these enormous power plants. The quality of the coal used as fuel at Lethabo is so poor that engineers jokingly refer to it as topsoil — about half of its content being ash. The power station dumps its waste gases straight into the atmosphere, discharging an annual 30 million tons of carbon dioxide (the major greenhouse gas) and 250 000 tons of sulphur dioxide (the major ingredient of acid rain). The laws of thermodynamics are such that power stations which convert steam energy into electricity are only 35 per cent efficient, at best. In addition, a vast amount of power is wasted by consumers. It is clear that coal-fired power stations are becoming anachronisms, and the sooner the world finds an alternative source of fuel, the better.

Waste is a major factor here. Prof Desmond Midgley, a retired University of the Witwatersrand hydrologist, said at a Cape Town conference on climatic change in 1989 that the power station complexes on the Transvaal coalfields produce twice as much waste heat as they do usable energy. 'By the year 2000 the heat release to the atmosphere from the Transvaal coalfields will probably be at the rate of 100 000 megawatts and that, apart from the accompanying sulphur dioxide, is bound to be environmentally significant.'

ABOVE
Pollution in the eastern Transvaal highveld

ESKOM, which produces two-thirds of the African continent's electricity, and which also supplies all of South Africa's neighbouring states, has a current capacity of 34 000 megawatts and is completing installations which will bring its capacity up to 43 000 megawatts. It could also have a further 5 000 megawatts if it utilized the Tugela River. Yet despite this massive capacity, the highest demand for electricity ever was only 21 800 megawatts, recorded in 1989. It is true that generating companies have to be optimistic: in depressed times they must prepare for buoyant times because it takes eight or nine years to construct a large power plant, and booms cannot wait for power utilities to catch up with them. But in view of the huge over-capacity in 1990 and the virtually unexplored possibilities of conserving power, one seriously wonders if South Africa needs to build another coal-fired power station. (See Chapter 7.)

As much as 85 per cent of South Africa's electricity is derived from coal and four-fifths of the country's electricity is generated in the eastern Transvaal Highveld. In the middle of the region's cluster of power stations are plants such as Secunda, Sasol's huge oil-from-coal complex which processes 43 million tons of coal a year, and several metallurgical industries. All of these have sprung from the coalfields and all help to put about 1,2 million tons of sulphur dioxide into the air each year.

Until 1990 the policy of the Department of National Health was to encourage industry to use high stacks (some up to 300 metres tall) in order to disperse gases and particles as widely as possible. In other words, the method was to use fresh air to dilute bad air. The tall stacks tended not so much to solve the problem as to spread it across the sky. For instance, before Sasol corrected the problem, its Secunda plant, 120 kilometres east of Johannesburg, used to release pungent hydrogen sulphide from its two 300 metre stacks. Under

WEEKLY MAIL

The pH of rain tested in the Highveld occasionally drops almost as low as that of vinegar

certain weather conditions, this invisible gas used to descend directly onto the Witwatersrand. One wonders what odourless, and therefore undetected — yet possibly dangerous — gases are being focused on this densely populated area.

The dispersal of gases and solids, even at 300 metres, is no permanent answer, especially on the Highveld, with its stagnant air. One sees demonstrations of this quite frequently when fly ash from ESKOM power stations meets an inversion 'roof' 500 metres above the ground and spreads out like a giant umbrella thorn tree, from which dust filters down, day and night. Although ESKOM's electrostatic precipitators are usually 98 per cent efficient, each stack still pushes into the air 2 000 tons of fly ash daily. When the precipitators break down, one sees huge clouds of thick smoke and dust literally tumbling out of the stacks and dropping directly to earth. The average fly ash fallout is at least 1 400 tons per day, according to official figures, which tend to be optimistic.

The public might breathe a little more easily if an independent agency, whose first commitment was to public health rather than to industry, were responsible for monitoring the air and were obliged to report annually on progress — or otherwise. Incredibly, the country's chief air pollution control officer is not obliged ever to produce an annual report.

As far as the smoky and sprawling sub-city of Soweto is concerned, smoke emissions frequently exceed the health levels recommended by such bodies as the United States' Environmental

'Don't oblige poor people to chuck out perfectly good coal-burning stoves and replace them with expensive electric stoves'

— Bill Muirhead

Protection Agency (EPA), whose standards seem to have been accepted by the Department of National Health. Nevertheless, until June 1990, when the MRC began its Birth to Ten survey in greater Johannesburg, the Department had conducted no meaningful studies on public health on the Highveld. The Departments of Agriculture and Forestry seem more concerned — they have at least begun to observe the effects of sulphur fallout on crops and timber plantations.

By the late 1980s, acid rain damage was becoming apparent in the eastern Transvaal's extensive pine forests. Trees, which grow in the earth longer than food crops do, tend to accumulate acid effects, but as the Transvaal's soil acidity is increased by the continued sulphur fallout, so one must also expect crop damage to begin manifesting itself.

The Tyson report noted that the pH of rain tested in the Highveld occasionally drops as low as 2,8 (similar to that of vinegar), and that the worst area for acid rain was the southern Transvaal around Standerton, which is an important maize and dairy district. The report predicted that acid rain would increase by another 25 per cent once ESKOM had all its turbines spinning. What then? ESKOM does not envisage a clean-up of power station gases, although it has announced that in future all new power stations will be designed with desulphurization equipment.

Car exhaust emissions also add to acid rain. When nitrogen oxides and unburned hydrocarbons produced from car exhaust pipes react with sulphur dioxide from coal-fired boilers, the result is a particularly acid brew. Acid pollution from cars is partly, and in some areas largely, responsible for destroying aquatic life and forests in Europe, North America and the Far East. Most of the acid damage in Athens and Rome, for instance, is as a result of car exhaust fumes rather than industry. In Britain, 30 per cent of acid rain arises from nitric acid (mostly produced by cars) and 70 per cent from sulphuric acid. In the last two or three years scientists have become increasingly concerned about what are called 'nox gases' — oxides of nitrogen which come from car exhausts and coal-fired power stations. They are now being linked with lung cancer and asthma by the American Lung Association and the British Lung Foundation. The latter believes that nox gases are more potentially dangerous than sulphur dioxide.

Most of South Africa's acid rain takes the form of a dry fallout. This is true even of a wet region such as Britain, where dry fallout is characteristic of urban areas and power station concentrations. Dry deposition does more damage than 'real' acid rain does.

As indicated above, it is just a matter of time before crop damage from acid fallout begins to appear in the Transvaal and Free State. Half of South Africa's high-production cropland — mostly comprising maize lands — is in the eastern Transvaal Highveld itself, according to Brian Huntley, Roy Siegfried and Clem Sunter, authors of *South African Environments into the Twenty-First Century*. According to a special report by the World Wide Fund for Nature

('Acid Rain and Air Pollution', 1988), if the annual average sulphur dioxide level exceeds 30 micrograms per cubic metre (or 30 parts per million), it will eventually result in 'a decline in the productivity of crops'. This figure is exceeded in most of the eastern Transvaal Highveld.

Britain's dramatic atmospheric clean-up was achieved not via smokeless stoves but via smokeless fuel

In 1990 ESKOM stated that it would cost R500 million to retrofit the two end boilers in each giant 'six pack' power plant to remove sulphur. The 'six pack' description stems from the fact that these plants each have six turbines, each equipped with a boiler. The boilers are placed in a row in such a way that the middle ones cannot be retrofitted with desulphurization equipment because there is no space. In other words, to reduce sulphur dioxide emissions by only one third would cost a total of R4 000 million for the larger power stations, and would increase electricity tariffs between 20 and 30 per cent. That was the prediction early in 1990. By June that year ESKOM was talking of R1 000 million to retrofit each power station and within a month of that, at an ESKOM workshop, the figure of R1 200 million was mentioned. One can be forgiven for thinking that ESKOM is guessing. It is likely that the figures are being inflated to discourage the public from demanding pollution controls. Many industries, not just in South Africa, use this ploy, usually ending with the warning, 'and it will be the consumer who pays in the end'.

Suggestions from fuel expert, Prof David Horsfall of the University of the Witwatersrand, that ESKOM should study the possibilities of partly desulphurizing coal by washing it, have been greeted with scepticism. Nevertheless, South Africa partly desulphurizes coal before shipping it to Japan. Horsfall believes that cleaning coal before it is burned would not only ensure a reduction in sulphur emissions from all power station boilers (instead of just the last two), but would also be considerably cheaper than the conventional method.

Bearing in mind the very high cost of desulphurization equipment, clean coal-fired power stations may soon cost more than nuclear plants. Coal plants cost less to build but their running costs, because of the enormous amount of coal that has to be mined and transported, are, in the medium term, greater than those of nuclear plants.

The Soweto dilemma

Ian McRea, ESKOM's chief executive, in confirming his organization's reluctance to embark on a desulphurization programme, argued that as far as air pollution was concerned, there could be only one priority: '70 per cent of South Africans do not have access to electricity in their homes,' he said in a broadcast in May 1990. 'We have to get electricity, as quickly as possible, to black urban areas such as Soweto.'

McRea estimated that this would cost R6 000 million and said it would have more impact on improving the quality of life for more people than cleaning up power stations. In any event, he said,

Until South Africa puts a realistic price on coal and electricity, there is little likelihood of any money being left over for much-needed coal research

Soweto was two-and-a-half times more polluted than the eastern Transvaal Highveld. Nobody would argue about the need for electricity in black areas. But McRea is wrong if he believes that this will solve the problem of smoke pollution.

A British delegate at a Soweto workshop on air pollution in June 1990, Bill Muirhead of Falkirk, said:

> Stoves don't smoke. It's coal that smokes. So, clean the coal! Whatever you do, don't oblige poor people to chuck out perfectly good coal-burning stoves and have to replace them with expensive electric stoves.

The point was reinforced at the same conference by Mike Harris of Richlab, a coal technology research centre in Johannesburg, who pointed out why an electric stove was no substitute for a coal stove. An electric stove merely cooked meals, he said. But a cast-iron coal stove not only cooked meals and kept stews simmering for hours while everybody was at work, but also heated all the household water, kept the house warm, kept the kettle boiling, boiled the washing and dried it overnight, and created a convivial environment as the family sat around it in the evening. Muirhead noted that when (soon after London's Killer Smog) England brought in smokeless zone laws, it did not ban the coal stove. 'There would have been riots if they'd tried.' He said that even today, coal stoves are favoured over electric stoves by large households as well as many middle-class homes in Britain.

Britain's dramatic atmospheric clean-up was achieved not via smokeless stoves but via smokeless fuel. Only when the option of buying smokeless fuel was open to all were smokeless zones declared. That, according to some, is the direction in which South Africa should be moving. South Africa is one of the very few places where smokeless stoves are still being made. While there is no export market for smokeless stoves, there would be a significant export market for smokeless fuel.

But until South Africa puts a realistic price on coal and electricity, there is little likelihood of any money being left over for much-needed coal research. In 1990 there was not one meaningful coal research project being conducted in South Africa. This was hardly surprising. The 1990 price for average quality coal was around R42 per ton at the pithead — practically the cheapest in the world. In view of the country's pollution problem, a resource economist might wonder at the sanity of this. The worst and smokiest kind of coal was selling at R11 a ton in 1990 — about R2,30 less than the 1985 price.

The Worldwatch Institute suggests that an electricity conservation programme — something ESKOM has never tried beyond advising householders that fluorescent lights use less electricity than ordinary bulbs — can mean the difference between building and not building a power station. If South Africa cut its demand and so saved building another megawattage station, which in 1991 terms would cost (with desulphurization equipment) R12 billion, then the saving could at least be used to pay for a significant reduction of sulphur dioxide in existing power stations.

The hole in the ozone

In 1987 an international meeting in Montreal called for a worldwide curb on chlorofluorocarbons (CFCs) which, only months before, had been identified as the cause of the holes in the ozone layer above the polar regions. Twenty-four nations signed immediately and several signed later, including South Africa (in January 1990). South Africa has undertaken to freeze CFC production at 1986 levels and to cut production by 20 per cent by 1992 and by a further 30 per cent by 1998. But the country's only manufacturer of CFCs, the chemical and explosives giant AECI, has stated that it will cut production by 50 per cent by 1992. (Aerosols use 50 per cent of their production and these are disappearing rapidly from shops.) South Africa is responsible for 1 per cent of the world's CFCs, Britain for 3 per cent and the United States for 25 per cent.

In twenty years of mounting anxiety over the deterioration of the human habitat, nothing has alarmed the world's politicians more than the sudden discovery that there was a gaping hole in the ozone above the Antarctic. The extent of it shook scientists too, especially as such an event had not been predicted in any of their computer models. A similar phenomenon occurs over the Arctic in the northern Spring. The damage already done will persist for decades unless there is a third revision of the international agreement aimed at actually repairing the damage.

Ozone is a form of oxygen but, unlike the oxygen molecule, which has two atoms (hence its designation as 0_2), ozone has three (0_3). As much as 90 per cent of the ozone in the atmosphere is high up in the stratosphere and for aeons, molecules of ozone, whose lifespan is a few weeks, have undergone a process of destruction and regeneration while maintaining a balance. If all the ozone were compressed into a layer at sea level it would be only 3 millimetres thick: in other words, there is very little ozone. There is enough, however, to absorb a critical amount of the sun's ultraviolet rays which, if the ozone shield did not exist, would lethally irradiate life on earth. One of the effects of the thinning of the ozone shield is that it will cause an increase in skin cancer, but a still greater threat is that it will aggravate global warming.

In twenty years of mounting anxiety over the deterioration of the human habitat, nothing has alarmed the world's politicians more than the sudden discovery that there was a gaping hole in the ozone above the Antarctic

The hole above the Antarctic was discovered by the British Antarctic scientist, Joe Farman, in 1982. Up until then some scientists, finding their readings on ozone levels were no longer making sense, blamed this on faulty instruments. When, in the spring of 1987, the hole was found to be 8 kilometres deep and as large as the United States, everybody was galvanized into action.

Joe Farman, quoted in the London *Sunday Times* in 1989, revealed that 'in one layer of the atmosphere (over Antarctica) we actually lost 95 per cent of the ozone in six weeks'. In other words, ozone which shielded the earth from ultraviolet rays was being neutralized. Finding the smoking gun was one thing, but who had been holding it? The culprit was a group of very stable manufactured compounds, CFCs, with widespread everyday uses, which were

WEEKLY MAIL

ABOVE
Johannesburg in the morning

ascending into the atmosphere, perhaps taking twenty-five or more years to reach the stratosphere, where the sun released their chlorine, which then destroyed the ozone. In the polar spring the first rays of the sun after the long sunless winter trigger this reaction. Afterwards the hole repairs itself. But just how big might this hole get? And exactly how much damage can the ozone layer tolerate before it becomes irreparable? Even if we immediately stop using CFCs, the damage is going to continue because of the years it takes for CFCs to climb through the atmosphere.

CFCs were invented in the late 1920s and today have many applications. In South Africa almost 50 per cent are used in aerosol spray cans, about 20 per cent in refrigerators and air conditioning units as coolant gases, and nearly 30 per cent are used to produce foam plastic for such items as fast food packaging and polystyrene cups. A small amount is used in the form of cleaning solvents in the local computer industry, but the computer industry is a major user overseas.

CFC-releasing foam plastic is already banned in some countries, mainly because it is light, bulky and non-biodegradable, and thus very expensive to dispose of in the waste stream. The biggest prob-

lem is finding a reasonably priced substitute for CFCs as coolants. In South Africa the gold mines (which have the largest refrigerators in the world in order to cool the air deep below ground) have vented tons of CFCs whenever they have overhauled their systems. Once it became clear that each molecule of chlorine from these CFCs was destroying 10 000 molecules of ozone, the mines began storing the discarded gases. They are waiting for the day when a technology will be developed to clean and re-use them. AECI is at present working towards developing such a technology. The mines use about 400 tons of CFCs a year — about 7 per cent of South Africa's total consumption.

When it comes to substitutes for CFCs, Third World countries are obviously not keen to face the extra expense involved. Their delegates at conferences have complained that it is unfair for the West, having benefited from years of using cheap CFCs — in South Africa they cost about R5 500 a ton — to now expect the Third World to forsake them. It is likely that the West will subsidize CFC substitutes for the Third World.

Two stopgap coolants will be hydrochlorofluorocarbons (HCFCs) and hydrofluorocarbons (HFCs): the latter contain no chlorine at all. However, both will continue to destroy ozone,

Even if we stop using CFCs immediately, the damage will continue because of the years it takes CFCs to climb up through the atmosphere

'What is frightening is that for every 1 per cent of ozone depletion it is predicted that there will be a correlating 2 per cent increase in skin cancers'

— Malcolm Scourfield

although at a slower rate; and they will be six to fifteen times more expensive than CFCs. There are less harmful substances, and these are being belatedly researched. There is talk, meanwhile, of replenishing ozone by attaching tanks of ozone to airliners for release at high cruising altitudes.

From a health point of view, aerologist Prof Malcolm Scourfield of Durban says: 'What is frightening is that for every 1 per cent of ozone depletion it is predicted that there will be a correlating 2 per cent increase in skin cancers.' Two years ago the World Health Organization predicted that a 10 per cent depletion in ozone would result in a 3 per cent increase in ordinary skin cancers and a 1 per cent increase in melanoma, the malignant version of skin cancer. Already, according to the MRC, 60 per cent of all whites over the age of sixty suffer skin cancer, mostly of the benign type. People with black skins do not suffer as much from exposure to ultraviolet light. What is not yet known, is what extra ultraviolet light will do to surface fish and plankton of the southern oceans, which provide massive amounts of protein, and what it might do to light-skinned livestock.

Ironically, despite fears regarding the seasonal thinning of high-altitude ozone, the other fear is that low-level ozone — which is an eye-smarting poison — is *increasing* over the middle latitudes of the northern hemisphere, because of industrialization. Gribbin estimates that ozone — produced when vehicle and factory fumes heat together in the sun — is thickening over industrial regions by between 0,5 and 4 per cent (in parts per billion) depending on the area. He says that at the beginning of the Industrial Revolution there was probably a quarter of the ground level ozone that exists now, and by 2030 we can expect today's level to double. Again nobody can predict what precisely will happen. Gribbin says low-altitude ozone, as an agent in the greenhouse effect, is a far greater problem than either nitrous oxide or methane.

The greenhouse effect

There is no doubt that if the existing great forests can be preserved, or rather conserved (for there is no reason why they should not be utilized on a managed basis), and if every country establishes a tree-growing programme, we could mop up a significant proportion of the world's excess carbon dioxide, which is the most important of the 'greenhouse' gases. Australia has already announced that it will plant a billion trees during the 1990s. This is enough to absorb 50 million tons of carbon dioxide, a fraction of 1 per cent of the world industrial output. South Africa, through the Jewish National Fund, is planting 5 million — not nearly enough for a country which is dumping so much carbon dioxide into the air, but it is a start. Five million trees would absorb 250 000 tons of carbon dioxide — less than one thousandth of South Africa's annual output.

South Africa is probably releasing about 250 to 300 million tons of carbon dioxide into the atmosphere annually — about 1,6 per cent of the world total. Low though this figure is, it is still the fourth or fifth highest in the world after the Soviet Union, the United States and Japan. There is likely to be an international law of the air before the 1990s are out, and countries which go to the expense of reducing carbon dioxide emissions may well boycott goods from those who refuse to do so.

While tree planting definitely helps, there are even more practical options for reducing carbon pollution:

- Stop building coal-fired power stations (unless some technological solution is found to trap carbon dioxide).
- Use more hydroelectric power.
- Design buildings so that they require less electricity.
- Develop alternative sources of energy such as solar and wind power.
- Use nuclear power. Power from nuclear fission may be something of a Faustian bargain, but it seems to be a sensible and less environmentally risky bridge between today's crude coal-fired plants and the benign fuel of the future.

The ideal power source will be something which is self-renewing and environmentally benign and which poses no undue risks to those working with it. This is where governments have failed us; they have been beguiled by science into shovelling money into nuclear fission research at the expense of the most logical power source of all — the sun. Nuclear scientists and those whose interests are vested in conventional fuels will no doubt produce all kinds of reasons as to why solar power is uneconomical. However, the world is fortunate in having the example of the United States with its open-minded attitude towards technology. Currently that country has several dozen solar power plants, each one becoming more sophisticated and promising than the last. In California an 80 megawatt plant is in operation, providing enough power for a town, and in the near future one of 450 megawatts should go into operation. California has the lead in this field because it has, in several regions, many hours of guaranteed sunshine. So does South Africa. Two-thirds of Israel's homes have solar panels which provide hot water and stored power for lights at night. In fact, the Kruger National Park's new camps use solar-powered batteries for lighting, and South Africa's railway system uses it for signals in remote areas, as does the freeway system for its roadside telephones.

A major reason why South African households do not utilize the country's abundant sunshine to at least heat their water is that the financial incentive is low. Electricity is relatively cheap (just consider how we waste it) and the cost of solar panels, because of the small demand, is relatively high. By the time a household is about to recoup its costs for solar power, the panels need replacing. If the government were to give tax rebates on solar power, there would be far greater incentive. Eskom has said that solar power will not save much electricity because domestic users consume so little compared with industry and commerce. This attitude, however, reveals a certain chauvinism. From a householder's point of view, using solar panels to heat domestic water alone would save one-third of

Second only to coal as a major threat to Earth's atmosphere is the motor vehicle and the fuel on which it runs

the household bill. Collectively, householders could reduce the national pollution load by 1 or 2 per cent.

The motor vehicle

Second only to coal as a major threat to Earth's atmosphere is the motor vehicle or, more specifically, the petrol and diesel fuel on which it runs. The world's 1,3 billion motor cars contribute a highly significant proportion of the more serious atmospheric pollutants. Every car is a mobile chemical factory discharging its toxins right where people live. Each year another 45 million cars are manufactured.

Just as Henry Ford is supposed to have said, 'You can have any colour car you want as long as it is black,' so tomorrow's ministers of health will be saying we can have any colour car we like so long as it is 'green'. I doubt that the petrol-driven engine will be taking us very far into the future. Indeed, Los Angeles has announced that it will phase out petrol engines entirely within the next fifteen years.

Scientists at the University of Los Angeles say that petrol and diesel fumes cause up to 30 000 premature deaths a year. In 1986, according to the WorldWatch Institute, 75 million Americans (almost one in three) lived in metropolitan regions where carbon monoxide, particles and ground level ozone from car exhaust gases failed to meet air quality standards. In industrialized countries cars contribute three-quarters of carbon monoxide emissions, half the nitrogen oxides, just less than half the hydrocarbons and about one-eighth of the particulates in the air.

The traffic-generated smogs for which places like Los Angeles are notorious can cut visibility down to 30 metres. In fact, these photochemical smogs are not really smogs in the strict sense of the word because they contain no natural fog and occur in dry air. The components of California's lung-tickling, eye-smarting smogs are entirely of human origin, being formed mostly from car exhaust gases suspended in stagnant air, where they heat together in the sun to become a highly reactive brew. Johannesburg and Pretoria experience similar but less severe photochemical smogs and Rand Airport, on the east side of greater Johannesburg, has several times been closed to light aircraft because of the photochemical haze which is often thickened by industrial smoke.

Cars also contribute a fifth of human-produced carbon dioxide, the major greenhouse gas.

It is unrealistic to believe the car can be ruled out of our lives, but it is just a matter of time before it changes in a fundamental way. The 'town car' as well as the buses of 2020 will almost certainly be silent and probably electrical. Cars will be smaller and slower than today's vehicles. They will probably be less powerful and only police and, possibly, emergency vehicles will be able to travel at speeds over 130 kilometres per hour. Maybe we will have town cars as well as long-distance 'touring cars', the latter being larger and more powerful. But in the cities, traffic will be so quiet that drivers will be able to talk to each other in traffic jams. (There will surely be traffic jams.)

At present, progress towards developing a truly advanced car is being hampered by a powerful lobby with a significant stake in the status quo. A friend who is developing a revolutionary battery-driven bus was asked by a manufacturer of utility vehicles: 'What are you trying to do to us? What will happen to all the people working in spark-plug plants or carburettor factories?' Imagine if Victorian horse traders had tried to stop the emergence of the horseless carriage on the grounds that farriers and blacksmiths would become jobless!

Practically all the work being done on developing a cleaner car is directed at modifying the engine or trapping the gases, instead of preventing the gases from forming in the first place. The oil lobby believes the answer to cleaning up air pollution produced by traffic can be found by staring down exhaust pipes. Many see the advent of the platinum exhaust filter as the final solution. In fact, the platinum-based catalytic converter which is now being made compulsory in some countries, does reduce hydrocarbon emissions by 87 per cent, carbon monoxide by 85 per cent and nitrogen oxide by two-thirds over the lifetime of a vehicle. However, this cannot be the final answer — for one thing, platinum itself is a finite resource. For another, oil is so limited in supply that several times in the last thirty years the world has run the risk of global war over a commodity for which there could be effective substitutes.

Oil is so limited in supply that several times in the last thirty years the world has run the risk of global war over a commodity for which there could be effective substitutes

There is no simple answer to South Africa's diverse air pollution problems — nor the global problems — but there is one essential move we have to make at national level. South Africa needs something like America's Environmental Protection Agency, an agency which determines environmental policy and standards and sees they are implemented. It has Cabinet status.

Currently no government department is responsible for safeguarding the environment against air pollution such as acid fallout and heavy metal contamination. The Department of Health is interested only in public health and, even then, it defends industry first and has never prosecuted any polluter. Nor does it employ lawyers. (By contrast the Department of Water Affairs employs ten lawyers and frequently prosecutes polluters.)

Like many government departments among whom various environmental responsibilities are scattered the Department of Health needs to feel the hot breath of the public via an environmental protection agency which is squarely on the public's side.

•JENNY MUFFORD•

To all intents and purposes, Jenny Mufford lives out in the country. The directions take you miles from smoggy Johannesburg, way beyond suburbia, onto a bone-shakingly rutted gravel road and up to the simple house on a small-holding. Perfect, really: fresh air, sunshine, plenty of space for the children ...?

Well, that's the theory. The reality is very different. Step out of the car and you are greeted not by the fragrances of nature but by a lingering chemical smell, by air that is almost tangibly dirty. The lowering clouds that cast a pall over the sky are not the work of nature but the product of Sasol and nearby Iscor, and of the hundreds of thousands of coal fires lit in the surrounding townships.

No wonder, then, that when she is asked what has motivated her single-minded battle against air pollution, Jenny Mufford answers simply and angrily, 'you can't breathe here'.

It is ironic in many ways that this woman who grew up on a farm and, despite stints in London and Johannesburg, considers herself a country person, should have spent the past eight years breathing the filthy air of what she calls the 'Vile Triangle'.

One of her earliest memories of being aware of pollution is her horror when, as a child, she first saw Sasolburg. 'I couldn't believe how ugly and filthy it seemed. I knew I could never live anywhere like that.' But now she does. A transfer for her husband left the couple with no choice.

With the birth of her first daughter, Mufford became aware of the toll the polluted air was taking. Winters were a nightmare. Helen, though resistant to most of the germs and viruses that attack children, had to sleep sitting up in an effort simply to breathe through what Mufford firmly believes to be the pollution-induced blocked nose and chest.

'I learnt physiotherapy techniques to make her life more bearable, but it doesn't make my life more bearable — it is a constant battle, the winter really comes down hard on us.'

The year that the whole family came down with a strain of hepatitis was the turning point for Mufford. 'I realized that my resistance was taking a knock and that this constant breathing in of all these poisonous gases over a period was having a terrible effect on all of us.'

A miscarriage, followed by the birth of her second daughter with a defective heart, was the final spur which prompted her to launch her campaign. 'My instincts told me that these things were caused by the air I was breathing and the water I was drinking, because what goes up goes down and gets into the water table'. However, Mufford acknowledges that she has no facts to establish a connection between the congenital problems and the atmosphere.

Encouraged by an earlier anti-air pollution drive in Vereeniging, which had attracted 10 000 signatures, and drawing on an American concept, she sent out a health questionnaire to householders in the surrounding areas. Taking with her the results of her survey — which revealed, she said, that a substantial majority of respondents suffered from respiration-related illness — she was one of those participating in a meeting between a group of concerned people and an official of the Department of National Health. Mufford claims that he 'treated us as though we were all ignoramuses. He didn't even take the signatures with him.'

The meeting might not have been a success, but the momentum had been established and Mufford was now on the campaign trail. Her next move was to join the National Association for Clean Air (NACA) and contribute a paper entitled 'Aspiration or Procrastination — A Resident's Perspective' to a conference in 1988.

But Mufford was impatient for results and felt

that NACA, which she sees as scientifically-based and industry-connected, did not move fast enough. Looking for support, she gathered around her a like-minded group of people, started the Anti-Pollution Appeal Committee (APAC), and set out to learn as much as she could.

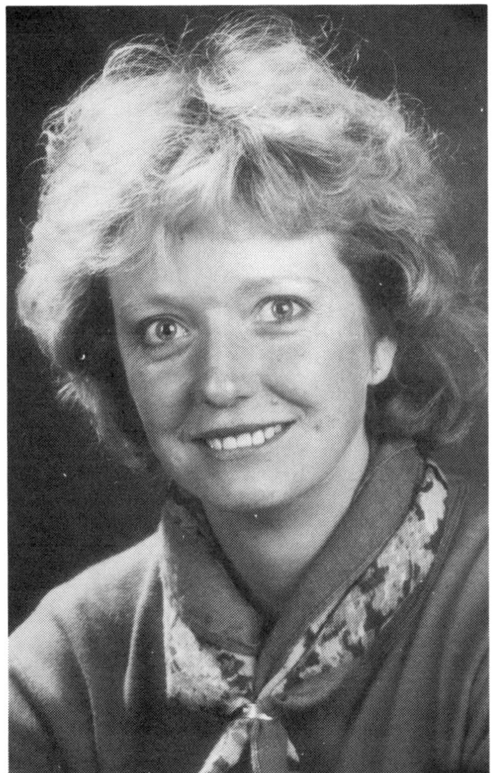

APAC's first move was to educate the public through a series of brochures and alert the press to the issues. With the formation of the committee, the pollution issue took over Mufford's life. 'I would,' she says wistfully, 'have liked to have been able to get a network going, which didn't happen.' What did happen, she says, was that her telephone rang constantly with threatening calls. 'We were labelled hysterical housewives and we didn't have a very good name. I started to get a bit paranoid,' she confesses.

One major success notched up by APAC was an art competition in which school children were invited to portray the way they saw the environment they lived in. What emerged was not a pretty picture. An exhibition of the children's art was held in conjunction with an exposition in which local industries were invited to demonstrate their contribution towards pollution control.

The problem was that none of the efforts seemed to make much difference. Industrial and governmental hackles were raised, but little impression was made on the pall of pollutants or on official resistance to action.

The Government, claims Mufford, still does not accept that high pollution levels affect health. They are, she says, planning an epidemiological study which will take ten years to complete and requires the cooperation of parents who are, apparently, reluctant to give it. But, she protests, 'all that information is here, it doesn't have to take ten years. In the US they did a broad-spectrum survey of over 3 000 people which took two months.'

Impatient, frustrated and bitter, Mufford saw her efforts crushed beneath the weight of industrial opposition and the slow-moving wheels of bureaucracy. APAC disintegrated and she has thrown in her lot with Earthlife Africa, joining her cause with the broader environmental issues that are tackled by that organization.

Despite the setbacks, the frustrations and a hatred of being labelled the 'air pollution lady', she is determined to fight on. 'There's a need for people to talk about pollution and I will carry on fighting for cleaner air until I believe my children don't suffer as they now do.'

She admits to feeling bitter. 'The evidence is there. We know that 2 000 lakes have died and that air pollution causes acid rain. It's not necessary to go on testing and lining other people's pockets, it's been done.

'We don't want to get to the stage where we have to spend billions cleaning up the air. We need to do something right now while we can still afford it. If we carry on like this we will all pay.' ❐

11

WASTING AWAY

South Africa and the global waste problem

◆ PETER LUKEY ◆ CHRIS ALBERTYN ◆ HENK COETZEE ◆

One of the activities that sets the human species apart from any other living organism on the planet is our practice of producing waste that not only threatens our own kind, but also endangers the entire natural world. In nature, substances created and discarded by living organisms become nutrients or building materials for others. This cyclic re-use of raw materials ensures the continuity of life by recycling finite resources: therefore no true waste exists in natural systems.

Waste is usually defined as material that has no value for sale, for productive use or for recycling. This way of thinking has led to the degradation of the entire biosphere through the pollution of land, water and air and is now threatening our very existence. The uni-directional use of resources represents a system that can obviously not sustain life indefinitely, owing to the finite quantity of resources available. Thus a new system is essential if we hope to see the survival of future generations of life on Earth.

'Sustainability' has become the watchword among people exploring green economics and technologies. The concept of sustainability is based on the important lessons learned from natural sustems and is now regarded by many environmentalists and a growing number of governments as the only principle capable of slowing our present rate of environmental destruction.

The creation of waste

It can be argued that all industrial endeavours create waste in one way or another. In many cases the mining or cultivation of raw materials required for industry leaves a veritable wasteland in its wake, while industrial activity all too often pollutes both water and the atmosphere. All this is done in the name of a consumeristic lifestyle based on the uni-directional use of resources.

Household waste

Most middle to upper middle class households in this country produce an endless stream of bags of waste. Johannesburg alone disposes of some 780 000 tons of household refuse a year, and this total is rising steadily. Unfortunately, this waste does not simply disappear. The amount of waste produced by an average household in a week multiplied by the fifty-two weeks of the year represents a garbage mountain that has to be disposed of somehow or other. Much of this waste is toxic in the sense that it presents a threat to biological life. It has been estimated that the average American household throws away 25 kilograms of hazardous waste per person per year. Methods of recycling — or, better still — re-using this waste already exist — but at present consumer ignorance about the threats to the environment inherent in the disposal of this waste, coupled with the apathy or profit motivation of those people who do understand the threat, ensures that these systems remain unused.

BELOW
A rubbish dump in Johannesburg

JUSTIN SHOLK

THE POOR SHALL INHERIT THE WASTE

The Pietermaritzburg toxic waste dumping saga of 1989 illustrates well the inefficiency of this country's environmental legislation. What laws there are — and there are none that pertain specifically to toxic waste dumping — are clearly biased in favour of maximizing profit rather than protecting the environment and the people who depend on it for sustenance.

More than 6 000 drums of toxic waste were dumped by a recycling concern in various places on the outskirts of Pietermaritzburg. As is the case all over the world, it is always the economically and politically least powerful sectors of society that face the dangerous consequences of waste dumping. The leaking drums containing organic and inorganic lead, cadmium, chromium and mercury landed on the doorsteps of squatters.

The businessman responsible for this had taken advantage of a legislative vacuum and had used market forces to put all opposition out of business. Because there was nothing to prevent him from dumping waste indiscriminately, he was able to offer industries the cheapest disposal option. He removed the waste and recycled it, albeit inefficiently, to retrieve solvents which were then re-sold.

When the Pietermaritzburg branch of Earthlife Africa exposed the dumping of the recycled toxic residues, the Pietermaritzburg City Council's first response was that the squatters in the area had to be removed. However, the businessman was subsequently told to dispose of the waste properly — but Earthlife caught him burying drums on a nearby farm.

When the drums were unearthed by the Department of Water Affairs, officials said they did not have the resources to remove such a large quantity of waste and the dumper would have to do it at his own expense and in his own time. According to the Department, the health risks inherent in inhaling this exposed waste were so high that the people living nearby had to be moved until the waste could be taken away. Thus the police were instructed by the Department of Water Affairs to see that the people were moved elsewhere. No alternative accommodation was offered, however, so the people did not move — yet more than eight months elapsed before the last of the waste was gone.

Near the fifth and last dump site found by Earthlife Africa, 126 empty toxic waste drums were discovered to be in use at a large squatter camp, where they were being used for domestic purposes such as catching rainwater from makeshift gutters. Again the authorities took months to address the problem and to this day there are still empty toxic waste drums at this residential site.

Blood samples taken from four people living very close to one of the waste dumps indicated early signs of lead poisoning. However, a definite diagnosis could not be accurately made as these people were also malnourished and had worms.

More than twelve months after the discovery, the major dumpsites were eventually cleaned up, but the processing of the waste continues unabated. Given the legal loopholes and the attitude of the authorities, who seemed to favour the businessman in this matter, charges were never laid. The Pietermaritzburg dumping case was never taken to court because of the legal costs involved. ❒

JUSTIN SHOLK

ABOVE
A woman scratches through a toxic waste disposal site near Katlehong

Industrial waste

Industry is the main producer of life-threatening toxic waste, as well as mountains of 'inert' waste, or waste which is apparently not toxic. Figures of between 6 000 and 100 000 tons have been quoted for the amount of waste material produced by South African industry per year that is so dangerous that it either has to be burnt or encased in concrete.

In the case of high-level nuclear waste, no suitable method has been found for its disposal and it remains toxic for thousands of years. As yet no accurate figures for the amount of this 'intractable' waste have been made public. Hundreds of thousands of tons of gaseous waste are produced by industry and our current methods of centralized energy generation.

Current methods of toxic waste disposal

A large percentage of the waste that industry disposes of is toxic. Direct skin contact, inhalation or ingestion of many of these substances can cause serious health problems, yet of the hundreds of toxic chemicals disposed of annually, very few have been fully tested to discover exactly what health risks they pose.

In the main, solid and liquid waste is either buried or burned or, in some cases, simply dumped on the land or into the sea and rivers. The more responsible producers of toxic waste usually pay professional waste-disposal companies to deal with it. Others choose a cheaper route — that of illegal dumping.

How the landfill industry is playing with fire

IT has been found that microbacteria attack and break down the refuse in landfills, acting as biological reactors in a process which creates methane gas. This gas has tremendous potential as an energy source, but it too has its problems.

In Kent, England, a woman suffered mild scorching and deafness after a methane explosion in her house, which was next to an infilled quarry from which a company was extracting gas. The blast apparently occurred when the company decided to switch off a new gas collector system temporarily, thus placing excessive strain on an older pump. The migrating gas was ignited by a spark from a washing machine.

This is one of many reported cases of accidents caused by the migration of gas from landfill sites. AECI currently pipe methane gas over a distance of 17 kilometres from the Robinson Deep landfill site, just south of Johannesburg, to their factory, where it is used as an energy source. Ironically, this alternative source of energy is used in the production of cyanide, an extremely toxic substance.

Landfilling

Waste disposal companies in this country usually bury waste in clay-lined landfill sites after it has been treated, depending on the type and toxicity of the waste. This treatment may include dilution of the toxic substances with inert waste, chemical neutralization or, in the case of extremely toxic substances, encapsulation in concrete. Landfilling is currently the most commonly used form of waste management, although it is a technique that is becoming more and more unpopular overseas, for the following reasons.

The loss of valuable land

Landfilling implies the loss of valuable land. Owing to the increasing quantities of this form of waste, huge tracts of land are involved. Abandoned landfill sites can be used only for playing fields or parklands — presuming that the waste has in fact been safely contained before the sites are rehabilitated.

The geographic concentration of toxic substances

Second, there is the problem of the geographic concentraction of toxic substances. In view of the quantities and the variable chemical composition of the materials handled, the possibility exists that the cocktail of toxics created by this process could form more dangerous substances than the originals. Landfills often release large quantities of flammable gases. Furthermore, it cannot be predicted for how many years a particular site will remain extremely hazardous.

Groundwater contamination

Although many safeguards exist at some landfill sites to minimize the hazard of groundwater pollution, there are increasing numbers of incidents of the failure of these safeguards and the related contamination of valuable groundwater reserves.

Incineration

Another form of waste disposal that is being used extensively in developed countries is the incineration of toxic substances. This method too has its drawbacks.

The creation of dangerous products of incomplete combustion

Many toxins have to be incinerated at specific temperatures. Any variance in these temperatures may lead to the creation of toxins even more dangerous than the original ones, and the possibility exists that these new toxins could enter the environment via the incinerator smokestack.

Toxic incinerator waste products

Though incineration reduces the volume of toxic materials considerably, the remaining incinerator ash remains toxic. This ash has then to be treated as a toxic material and is thus landfilled. This obviously entails once again the problems associated with landfilling.

ABOVE

Rotting waste in the streets of Cape Town when 2 000 municipal workers went on strike in June 1989

Additional problems

There is a further hazard associated with both forms of waste management, namely, getting the toxic materials from the source to the waste handlers.

Transportation hazards

As a result of the tendency to centralize waste disposal facilities, thousands of tons of toxins are transported by land, sea or air to these sites. The possibility of serious accidental spills of these toxins into the environment increases as the quantity of waste grows. The fact that these substances must be handled twice increases the chance of accidental human and environmental contamination.

Energy waste

Another very important consequence of both these systems is that valuable resources are lost and valuable energy is wasted in replacing them.

The planet has limited supplies of non-renewable resources and the current system of resource utilization means that the resources present in the waste material are lost to us. Huge amounts of energy are used to replace them, resulting in a further loss of non-renewable resources and an increase in the pollution inherent in our current means of energy generation — a vicious circle?

The situation could rather be described as a downward spiral, because we value our present lifestyles above the needs of future generations.

Learning from nature: how to approach waste management

It is obvious that our current methods of waste management pose a

The tightening up of regulations in many countries has led to a situation in which it is more expensive to dump than it is to change to environment-friendly waste reduction

direct threat to the environment, waste valuable resources and lead indirectly to further pollution. These systems can never be solutions to the problem of waste as they address only the symptoms — namely, the growing mountains of life-threatening waste — and not the cause of the problem — the production of the waste.

Natural systems on earth provide easily-interpreted instructions on how to solve waste problems. These systems have been tried and tested for millennia and the survival of life on this planet has been the proof of their effectiveness. Natural systems use non-harmful products; they are highly efficient; and they involve recycling and recovery. How these models can be adapted to our present dilemma can be summarized as follows:

Product substitution

Waste produced in nature is not harmful to the environment in the same way that the waste humans produce is. Hence if we substitute products which damage the environment with non-polluting products, or use non-polluting materials in the production process, we can see significant reductions in waste creation.

Process modification

Industrial and other systems could be much more efficient in terms of usage of resources, waste production and so on. Simple principles of good housekeeping, in terms of machinery maintenance and the like, can reduce waste. The modification of existing industrial processes in order to produce less waste is also possible.

Recycling and recovery

Toxic materials used in production processes should be be recovered and recycled.

Elimination of dangerous products

What cannot be recycled, or is too dangerous to handle, should not be produced in the first place.

The introduction of these concepts into industry reduces waste and therefore reduces the negative impact of production on the environment. The systems described here mirror natural systems and, as such, are assured of success. It is for this reason that the global environmental movement is campaigning for their introduction.

Action by authorities and some responses to it

Legislation

The tightening up of waste handling regulations in many countries has led to a situation in which it is more expensive to dump than it is to change to environment-friendly waste reduction. This, coupled with strong financial deterrents in the form of fines for illegal dumping, has prompted many overseas companies and local authorities to go the green route. But this is not enough. If the legislation is not properly implemented and the situation is inadequately policed, there is nothing to prevent companies from simply ignoring

AFRICA'S REACTION TO THE INTERNATIONAL TOXIC WASTE TRADE: SOME EXAMPLES

AT the May 1988 meeting of the OAU, Nigerian President, Ibrahim Babangida, argued: 'no government, no matter what the financial inducement, has the right to mortgage the destiny of future generations of African children'. The meeting passed a resolution against dumping after calling the practice 'a crime against Africa and all its people'.

The Ivory Coast has passed legislation that provides for up to twenty years' imprisonment, and fines of up to 1,6 million US dollars, for involvement in toxic waste dumping.

A Nigerian official, Iuro Onabule, warned that his country was prescribing the death penalty for anyone involved in toxic waste dumping. He added: 'there will be no concessions to appeals from foreign governments.'

Zimbabwe's Labour Minister, John Nkomo, President of the three-week annual assembly of the International Labour Organization (ILO), spoke against waste trade and encouraged labour interests to involve themselves in the issue. Nkomo deplored the fact that industrialized countries were turning Asia, Africa and Latin America into chemical waste dumps. He urged delegates to the ILO to adopt international labour standards, and develop legislation on the use, handling and disposal of chemicals in order to protect workers who came into contact with harmful materials.

A statement by the Economic Community of West African States in 1988 read as follows: 'We cannot accept that at a time when industrialized nations refuse to buy our commodities at reasonable prices, these same countries are selling us death for ourselves and our children.' ❐

these laws. It is a fact that in countries where companies have changed their waste disposal policies, public awareness, state monitoring and policing and the actions of environmental groups have led to that change.

South Africa has a long way to go before a similar situation is reached and green industry becomes a viable proposition. Our current system relies heavily — and dangerously — on the integrity of industrialists, who bear the onus of disposing of hazardous waste correctly. This is largely a result of a lack of public awareness, a lack of adequate monitoring systems, and the fact that our laws and the penalties for their contravention are so lax. An unfortunate by-product of the current situation is that our lax environmental laws could act as an incentive for 'dirty' international companies to stay in this country, or reinvest here, since they provide foreign exchange and stimulate local economic growth.

Our current system relies heavily — and dangerously — on the integrity of industrialists, who bear the onus of disposing of hazardous waste correctly

Shape up or ship out?

Though the tightening up of waste legislation has persuaded some industries to 'shape up' environmentally, many have chosen another path — that of 'shipping out'. The incidents of exportation of toxic wastes have been widely publicized, yet this threat to the environment continues to grow. So-called 'developing countries' are the usual targets for these shiploads of toxic materials, generally for reasons of simple financial expediency. The developing countries often lack strictly legislated controls over the handling of toxic wastes, with the result that for the company doing the dumping,

WHAT CONSTITUTES TOXIC WASTE?

ACCORDING to the Basel Convention, which governs the cross-border movements of dangerous waste products, hazardous waste consists of products which are:

- explosive
- flammable (liquid or solid)
- liable to spontaneous combustion under normal conditions encountered during transport or disposal
- capable of emitting flammable gases on contact with water
- oxidizing (capable of giving off oxygen so as to cause or contribute to the combustion of other materials)
- organic peroxides (thermally unstable and capable of emitting large amounts of heat during decomposition)
- poisonous (causing acute damage to health, or even death if swallowed, inhaled, or in contact with the skin)
- infectious
- corrosive
- capable of emitting poisonous gases on contact with air or water
- toxic (causing chronic or delayed effects on health if they are inhaled, swallowed, or in contact with the skin — this includes carcinogenic substances)
- ecotoxic (toxic to the environment or ecosystem)
- capable or producing any substance after disposal which displays any of the above characteristics

the operation is far cheaper than it would be in the producer country. Furthermore, many developing countries are finding it increasingly difficult to repay the interest on foreign loans (let alone the loans themselves) received from the very countries producing the waste. Thus the foreign currency generated by accepting waste is a simple way of relieving the debt burden.

This state of affairs, whereby the producer country gets rid of its waste with the one hand and receives the money for that waste with the other, has been referred to as 'toxic colonialism', or 'garbage imperialism', by many African leaders. Sam Omatseye, writing in the Nigerian magazine, *African Concord*, expressed this outrage when he noted that the toxic waste trade 're-echoes what Europe has always thought of Africa: wasteland. And the people who live there, waste beings.'

Apart from the obvious risk to the people and environments of developing countries, the long-distance, sea-borne transportation of some of the most toxic substance known to humankind is a huge risk to the health of marine environments. Many cases of the unscrupulous dumping of waste have been exposed and this, coupled with public pressure, has led to the convening of many international conferences on the subject.

The Basel Convention

The Basel Convention, signed by thirty-four countries on 22 March 1989, constitutes an important step forward in controlling the international trade in toxic waste. At the time of writing South Africa had ratified the convention preparatory to signing it.

Environmental groups around the world hoped that the convention would ban the cross-boundary trade of waste; unfortunately, however, it did not go this far. According to Greenpeace, 'industrialized countries had the power to stop waste exports to the Third World; instead, they opted to legalize them.' Thus the most important mechanism of the convention was that exporters would be required to notify the governments of any exporting and importing countries before shipment took place, and would need permission before proceeding. In effect, this notification system has not restricted the international waste trade, it has simply set up a global tracking system for waste. A further provision requires all signatories to institute industrial processes which minimize the production of hazardous wastes and to develop adequate facilities to handle their own and imported waste.

While the convention may look as though it solves many of the problems associated with waste imports, and controls potentially dangerous practices, it is seen by many environmentalists as simply legalizing the existing toxic waste trade.

The main motivation for industries to export waste is that of cutting the costs of disposal. This almost always means that the environment pays the price, and the cheaper labour in Third World countries is usually exploited as well.

In addition, the transport of hazardous substances is in itself an extremely hazardous process. Licensing and legitimizing it, while it may close the door on irresponsible operators, still does not remove the dangers inherent in the transport process: trucks can crash and ships can sink.

For South Africa, the Basel convention presents new challenges. In terms of the convention, the importing nation must have adequate facilities and infrastructure to handle the waste. The South African government intends signing the convention, and it is possible that many of the new environmental regulations being proposed are intended to fill in the gaps that exist in the current legislative controls.

South Africa is in a better position than most other Third World countries to handle waste adequately, having a more developed infrastructure and greater technological ability. But South Africa has in common with the rest of the Third World a poorly developed system of inspection, and an environmental movement still in its embryonic stage. Thus the convention, while aiming to prevent the

THOR CHEMICALS IN THE NEWS

AN extract from an article in the *Washington Times* of 14 July 1988 records the response of Thor Chemicals, a company based in Cato Ridge, Natal, to claims by Greenpeace that the company was importing toxic waste:

John Dyer of Thor Chemicals said yesterday: 'We do not regard ourselves as importers of toxic waste. We supply American Cyanamid with mercury-based catalysts and, as part of the deal, we take back spent waste that is left over after they have used the product. We regard that as our responsibility', he said.

'This is done with the full knowledge of the American Environmental Protection Agency and with the South African Department of the Environment. Our facilities are inspected by the department, and each time there is a shipment, we notify them and they visit the plant', Mr Dyer said.

'We recover mercury from the waste. Our plant is environmentally safe and meets all the necessary standards. All the local authorities are satisfied that it is safe', he said. 'We believe that our approach is absolutely responsible and we go to great pains to ensure it is safe.'

AN extract from an article in the *St Louis Post-Dispatch*, 13 April 1990.

Thor Chemicals was experiencing problems with control of mercury effluent at the plant owing to heavy rains and also because of increased production. We have therefore advised them to suspend all operations, said Lin Gravelet-Blondin, assistant director of the Department of Water Affairs.

The Department said that it was sending experts to investigate the damage from mercury poison in a valley inhabited by Zulu people. The closing was expected to last at least a month, or until levels of contaminated water drop, authorities said.

The shutdown was ordered four months after an investigation by the *Post-Dispatch* revealed mercury contamination in a stream to be hundreds of times what is considered toxic. The stream, which runs behind Thor in the Valley of a Thousand Hills, is used for drinking, bathing and swimming ❐

destruction of the environment that has been seen in other parts of the Third World, could easily turn countries like South Africa into the First World's garbage bin.

ABOVE
Debris from Thor Chemicals found near Cato Ridge in Durban in 1989

Shipping out: Thor Chemicals

Late in 1989, journalists discovered that the Thor Chemicals plant in Cato Ridge near Durban was releasing large amounts of mercury into the nearly Umgcweni River. This river, which eventually feeds the Inanda Dam, the source of Durban's drinking water, was found to contain some of the highest concentrations of mercury ever recorded. The mercury was released by a recycling plant, which processes imported toxic waste.

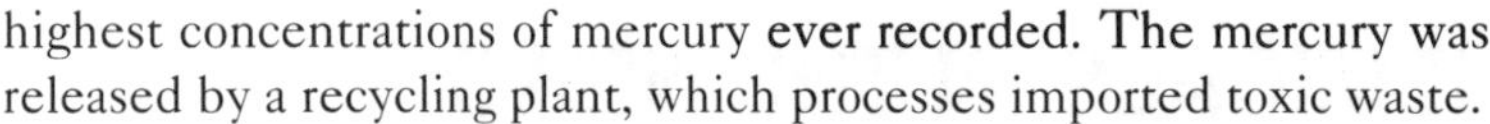

Thor state that the mercury recovery plant responsible for the pollution recycles mercury catalysts sold to overseas markets by the company. They and the South African government therefore regard the mercury waste that they import as a raw material. All in all, ten tons of mercury waste are imported by Thor from the United States company American Cyanamid every year, and Greenpeace have discovered evidence of contamination in some of this 'raw material'.

In 1990 Earthlife Africa's Pietermaritzburg Branch launched a protest campaign against these waste imports, with support from other Earthlife Branches, Greenpeace and the COSATU-affiliated Chemical Workers' Industrial Union (CWIU). As a result of this national and international outcry, the Department of Water Affairs suspended all operations involving mercury at Thor, although these were allowed to continue after Thor rectified design and construction faults in the plant. Soon afterwards, in September 1990, the South African government banned the importation of toxic waste.

This ban is, however, full of loopholes and doublespeak. The Departments of Health, Environment Affairs and Water Affairs have indicated that they have no objections to the continued importation of mercury waste by Thor. The then Minister of the Environment, Gert Kotze, justified this stand, defending Thor Chemicals' right to import waste because the company was a bona fide 'manufacturing industry'. Although the ministerial proclamation banning waste imports clearly states that 'no hazardous waste may be imported', Koos Stander, the Department of the Environment's Director for Hazardous Waste, has said that other companies could be granted the right to import toxic waste if they intend to extract raw materials from it and if they comply with South Africa's water pollution and industrial safety regulations.

Undemocratic, corrupt governments and severe cash shortages make the so-called independent homelands ideal targets for toxic waste merchants

Shipping toxic waste to the 'homelands'

Undemocratic, corrupt governments and severe cash shortages make the so-called independent homelands ideal targets for toxic waste merchants. Ciskei abandoned plans to import large consignments of toxic waste after news of this scheme was greeted with a public outcry. The South African government played a major role in preventing the'homeland's' plan to build a port and incinerator to handle thousands of tons of waste from Italy and Germany.

A letter from the Transkei Development Corporation was quoted in the local media early in 1990: 'We have a wealth of ravines and canyons which could become useful territory for reclaiming if they were filled with sanitarily processed waste on which perhaps trees could be planted.' This 'sanitarily processed waste' has been described by Greenpeace as 'unknown quantities of unknown chemicals, presenting health threats of unknown magnitude and duration to the people and ecosystems of neighbouring communities'. Luckily, it appears as if Transkei's military ruler, General Bantu Holomisa, has refused to accept the deal.

In October 1990, Swiss arms dealer Arnold Kuenzler approached the South African consulate's scientific division in Switzerland with incineration and disposal plans for the Ciskei, Bophuthatswana, Transkei and Lesotho. Sources at the Foundation for Research and Development at the Council for Scientific and Industrial Research have confirmed that four 'homelands' have already received lucrative offers from overseas concerns.

The possibility of the densely-populated 'homelands' receiving toxic waste is frightening. Labels on drums will mean nothing to the majority of local people who are illiterate, and the possible results of an accident can only be imagined. Toxic waste exports only make sense to countries where labour is cheaper and environmental controls are weaker than in the exporting nation. Cheap, poorly-educated labour, weak controls over environmental matters, and the inefficient and often corrupt 'homelands' administrations, add to this recipe for a toxic disaster.

BELOW
Some of the debris found near Cato Ridge

The banning of toxic waste imports

A more effective way of stopping this form of Third World exploitation is the outright banning of toxic waste imports by target countries. In response to the recent explosion of international waste trade schemes, at least eighty-one countries, including South Africa, have banned the import of foreign wastes. In addition, major intergovernmental organizations like the Organization of African Unity (OAU) and the association of ACP countries (former European colonies in Africa, the Caribbean and the Pacific) have called for a ban on the export of wastes from industrialized to developing countries.

The South African government itself has lately reaffirmed its stand that it will not accept toxic waste from other countries, but the threat still remains. Waste traders have found many loopholes to pass off waste as raw materials, or as materials to be recycled by the receiving country. (These principles are well illustrated in the Thor Chemicals episode.) Of course, another major problem is the fact that South Africa produces toxic waste.

Relocating to 'Toxic Town'

Given that the toxic sluice gates in industrialized nations are closing, the waste problem facing the First World is becoming more and more critical. Dumping space is at a premium, and pollution control and prevention laws are becoming stricter. Western industrialists are seeking not only dumping sites for waste, but also places in which to locate the 'dirty' industries themselves.

The next five years could be crucial in determining directions in global industrial development and pollution prevention. A major factor in this equation is the unification of Europe scheduled for the end of 1992. Strict unitary environmental legislation is going to put the squeeze on those industries which pose a threat to the environment and to human health. Thus the 'dirty' industries may now resort to relocating to other countries that are not as far-sighted as those in Europe.

South Africa could be a prime relocation target for dirty industries, given the need for foreign investment as a solution to our economic problems. The local establishment of the 'dirtiest' industries that service the First World may be seen as a short-term solution, and as a long-term mortgage on this country's future.

South Africa has clearly found this prospect tempting and major feasibility studies have already been conducted. One cartel, financed by European concerns, has already spent R750 000 on a feasibility study which has been submitted to the government. The major aspect of this proposal is to build a new harbour near Port Nolloth on the West Coast, to import vast quantities of foreign waste there and burn it. The other significant part of this proposal is the establishment of a 'chemical city' around the harbour. Envisaged is a collection of the world's dirtiest industrial installations that could export their products to the West. Disposal problems would be 'solved' by the presence of an incinerator. ESKOM has also been conducting feasibility studies for the siting of its second nuclear power station in the area, guaranteeing a source of energy to these industries.

Seen in relation to the strict regulations enforced in First World countries, the saving on pollution control, waste disposal, occupational safety, and labour costs would mean that healthy profits could be made in the manufacture of dangerous chemicals. South Africa

JUSTIN SHOLK

ABOVE
Gloves sunk in toxic waste at a dump in Ulana Park near Germiston

could even offer to dispose of the wastes generated when these exported chemicals and their by-products are used in the First World. This may sound like science fiction, but it is happening already.

This state of affairs, whereby the producer country gets rid of its waste with the one hand and receives the money for that waste with the other, has been referred to as 'toxic colonialism' or 'garbage imperialism'

Individual action to stop the toxic time bomb

As individuals, we can help to stop what has been referred to as the ticking of the toxic time-bomb. Much can be achieved by lobbying political organizations as well as supporting public protest, but we can all help at a personal level as well.

In the household

The careful selection of products that are not over-packaged and the separation of household waste, including tin, paper, plastic and paper, for recycling, will greatly reduce our contribution to the waste mountain. The decreased use of toxic household products in itself decreases the environmental impact of toxic waste. Inform yourself about how your lifestyle impacts on the environment.

At the workplace

Actively campaign for a safer, waste-free workplace.

In the community

Make yourself aware of the issues and find out what is happening in your community. Report any dubious practices to the authorities and as well as to an environment group in your area. Join an environmental group, or if none exists in your area, start one of your own.

JUSTIN SHOLK

ABOVE
A blood bag and other medical waste which was dumped near Katlehong on the East Rand

JUSTIN SHOLK

ABOVE
A cow grazes on a rubbish dump where toxic waste was found near Katlehong township

•BEV GEACH•

Conservation, Bev Geach was told, is no job for a woman. So she became a microbiologist and, later, became involved in medical biochemistry. But her concern for the environment did not fade, and when she moved from Cape Town to Johannesburg and found herself unable to escape the pollution by taking a walk on a beach or up a mountain, she began to feel a need to get involved.

By then a mother of two, Geach spent what spare time she had reading and cutting out articles on environmental subjects. 'Then I saw a couple of articles giving statistics about things like the decline of the rhino population in the past ten years. That isn't such a long time. I wondered what I had been doing for the past ten years.' It was then that Geach decided she could not again work in any field that was not directly related to the environment.

During her initial involvement with Earthlife Africa, she discovered how hard it was to set up links with other involved organizations and individuals and started compiling a directory. And thus *The Green Pages* was born. Before the first issue was out, the concept had broadened to include not only the directory, but also articles and a diary of environmental events.

Geach's intention was twofold — to forge links between different environmental groupings and to interest them in thinking about broader issues. She wanted them to look at the environment in a more holistic way, taking cognisance of each other's work as well as of factors which at a surface glance might not seem to be environmentally relevant. She is passionate about the need to look at the whole picture — political, social, economic and environmental.

'A lot of the problem in this country,' she contends, 'is that people are trying to conserve furiously without thinking how the increasing urban populations and illiteracy are affecting the whole thing. People should be made more aware of the bigger picture. I don't think they are aware of the implications of having generations of children growing up badly educated, badly fed, surrounded by violence.

'We will never solve the environmental problems until we have dealt with the socioeconomic problems. If you can't care about other human beings, how are you going to care about the air, the plants, and the animals?

The trouble is, she points out, that by the time the socio-economic problems have been sorted out, it might be too late to clean up. She blames what she sees as a 'very male ethic' for much of the sorry state of the world and believes the search for solutions has reflected 'not so much a caring thing as a desire to control and dominate.' She cites, by way of example, the use of pesticides and the race for armaments.

'From a scientific point of view we try to look at everything in little parcels without seeing how it affects the whole picture until it's too late. Thinking doesn't go further than that problem.'

Geach struggles with priorities. 'On a global scale the greenhouse effect and the ozone layer are major problems, but in the South African context there are so many other major problems — soil erosion, pollution, air, water, marine life — one doesn't know which to tackle first.'

It all sounds overwhelming and Bev Geach admits to sometimes feeling incapacitated by the sheer magnitude of it all. Yet, she believes that something can and must be done and since she is here, she intends to be one of those who make a contribution.

'If I thought there was absolutely no hope of anything coming right I'd go off somewhere, and if I didn't have children I probably wouldn't get involved to the same extent. Having children, you have to think further than your own lifetime. Also, I think there is a chance that we can actually work things out.'

Among the broad issues Geach believes it is possible to tackle effectively is the level of pollution. 'A lot of pollution problems are related to politics, overcrowding, squatting, education; we *can* do something more and I think we should.'

She is also a great believer in individuals making changes at a personal level, in their homes and in their lifestyles. The world is, after all, she points out, full of individuals 'doing damage or making the decisions to do good. It's a question of growing awareness. 'I don't think one can change one's lifestyle immediately — one does what one can.'

In her own home, she tries to buy things in returnable bottles, take newspapers for recycling, limit the number of 'throw-away things' she brings into the home, and to return containers (such as polystyrene trays) to where they can be re-used. She has considered solar energy, but the initial costs are daunting.

However willing we are, though, the problems cannot be solved entirely by individuals. There has to be public sector intervention. An important example is transport. An effective public transport system would radically reduce levels of pollution from private cars. Geach believes that 'it is the role of the government to set up a social system and structure which is helpful. You need everybody cooperating.'

One's mind-set is crucially important here. 'We're spoilt in the First World section of this country, and terribly wasteful,' she points out. 'It's too easy to blame all the problems on overpopulation.'

The fault, she believes, lies in the fact that 'the balances are wrong — 20 per cent of the world's population is using 80 per cent of the resources.'

She sees two sides to the problem facing the planet. One is the First World side — over-consumption and the throw-away society — and the other the Third World side — deforestation and erosion. 'Both sides need to make concessions and adapt.'

There are times when Geach despairs and wonders whether all the prophets who forecast doom at the end of the century weren't right after all. There are other times when she finds she can still hope.

'There is a lot of increased awareness. I think there's a lag period before people who have become aware actually do something, but I think it's coming. It's a snowball effect and I think once one starts doing more good than harm that snowball effect will grow.'

It has to. 'If we don't start addressing it now, in ten years' time it will be too late.' ❐

12

FROM SOIL EROSION TO SUSTAINABILITY

Land use in South Africa

◆ DAVID COOPER ◆

Land degradation: a national and international perspective

'A great deal of land degradation in many cultures arises from unjust, inequitable access to land ... The capacity of the 'homeland' areas to sustain a given population has been far, far surpassed, arising clearly from the system of apartheid. On the other hand, the technically advanced agriculture of the white farmers also has inherent problems ... It is almost primitive in terms of environmental awareness.'— *Stan Sangweni, ANC environmentalist*

Worldwide, soil erosion is a serious and escalating problem. In the Third World, erosion is associated with and caused by poverty: rural people, often pushed onto marginal lands for commercial or political motives, are forced to strip the land of wood for fuel, and resort to overgrazing or overcultivation to make a living. Thus a cycle of land degradation and deepening poverty is set up. In the modern agricultures of the First World, on the other hand, monocropping and the overuse of agrochemicals have also led to soil degradation. Although it is not as severe or immediate as erosion in the Third World, the extensive nature of this degradation has helped to make soil erosion one of the most serious global environmental issues, threatening long-term security for millions of people.

Land reform is crucial to the creation of a new, democratic South Africa. Unequal access to land affects some 15 million South Africans who live in the 'homelands' or as tenants or labourers on white farms. Generations of apartheid and careless farming methods have created a near-wasteland in the 'homelands' and have severely depleted the resources of other rural areas. A new South African government must examine not only how land can be more equitably distributed, but also how that land can be used to sustain the rural population in the long term. Such a transformation requires a new land policy — one that ensures environmentally-sound land use and offers incentives for the majority of rural producers.

Soil erosion in the 'homelands'

For rural people, soil erosion is probably as serious a problem as unemployment. Both drastically affect the viability of the family or community. However, while families will readily identify unemployment as a problem, they do not consider the problem of soil erosion to be equally serious. Other aspects of the land question are more likely to warrant attention — land shortage, shortage of equipment and drought, for example. Once the topic is brought up for discussion, however, people begin to realize that erosion poses problems. They may identify its effects by saying, for example, 'Yes, the donga at the bottom of my field does affect how much food I produce,' or, 'The bare hills mean there is no grazing for my cattle'. But soil erosion is nevertheless ultimately perceived as a natural process rather than as a socially generated problem.

Is soil erosion a natural phenomenon? Certainly there are natural processes involved — wind and water remove soil, and raindrops strike the surface of the soil and dislodge soil particles. However, soil erosion also has social and political components: it is

> a result of underdevelopment (of poverty, inequality and exploitation), a symptom of underdevelopment, and a cause of underdevelopment (contributing to a failure to produce, invest and improve productivity) ... it can contribute to chronic food shortages, so-called natural disasters and hazards, drought, landslides, floods and undermine an entire country's development effort.' (Blaikie, 1985:9).

In South Africa, soil erosion is most serious in the 'homelands'. Clearly apartheid, as a form of social engineering, has forced millions of people to live under crowded conditions in sometimes environmentally fragile areas. The consequences have been devastating, taking the form of erosion both in the western parts where wind erosion has been most serious, and in the wetter eastern parts of the country, where water erosion and dongas are the commonest manifestations of the problem.

Herschel, on the boundary of Lesotho, is a typical example. Dongas 10 metres deep cut through the arable lands, contour banks have been cut away and the fields have lost tons of topsoil. Although erosion rates have not been measured, they must far exceed the 20 tons of topsoil for each ton of grain produced, quoted as an average for South Africa. In Msinga, 'old fields have completely washed away, opening up extraordinary expanses of stone' (Durning, 1990:8-9). In 1980, the Ciskei Commission reported that 46 per cent of land in that 'homeland' was moderately or severely eroded, and 39 per cent of the veld overgrazed. In the drier, western 'homelands' like Bophuthatswana and parts of Lebowa, overgrazing has ruined grasslands and led to severe bush encroachment.

'The bare hills mean there is no grazing for my cattle'

— Farmer

People forced to subsist in rural areas without access to adequate resources and infrastructure have no choice but to strip the environment in an attempt to survive

Alan Durning, an analyst at the Washington-based WorldWatch Institute, gives four interlocking causes of soil erosion in the 'homelands', all linked to apartheid: poor land, politically enforced overcrowding, labour shortage and poverty.

Durning's claim that all the 'homelands' are located on poor land is not altogether correct. Although the 'homelands' have slightly worse land than the average arable landholding in South Africa, substantial parts of the 'homelands' are in the wetter, eastern half of the country. While much of this land is fertile, it is also fragile, in the sense of being vulnerable to soil erosion. The soils are easily eroded if ploughed, and much of the land slopes steeply. It would in fact be more accurate to say that the cause of erosion has been inappropriate use of the land. Overcrowding has forced the population to try to grow food on land better suited to pastoral production. While it is common to blame poor farming methods for rapid soil erosion, the methods are themselves only a symptom of the problem. Identifying the underlying root causes of soil erosion gives a more accurate picture.

Durning's second factor, overcrowding, is undoubtedly the major reason why erosion occurs. Apartheid policies have forced millions of people into rural 'homelands'. People forced to subsist in rural areas, without access to adequate resources and infrastructure, have no choice but to strip the environment in an attempt to survive. This is a worldwide phenomenon. Population densities in the 'homelands' are ten times those in the rest of rural South Africa. QwaQwa, the most crowded homeland, has a population exceeding 500 000 — South African government planners believed the optimum population should be 25 000.

Overcrowding has been made worse by unequal access to land in the 'homelands'. Even though average incomes for these areas are grossly unequal compared with average white South African incomes, there are also differences within the homeland populations themselves. Up to 80 per cent of all farm produce comes from 20 per cent of producers, according to the Development Bank of Southern Africa (DBSA). In some areas, three or four landholders own 80 per cent of the livestock being grazed on communal land. There is also widespread landlessness in the 'homelands'. Even in the Transkei, which has the most viable land, fewer than 50 per cent of villagers are allocated a field, and over 60 per cent have no cattle. Thus the belief that living in rural areas gives people access to a means of livelihood is a myth — overcrowding in the 'homelands' is simply too severe for that.

Durning's third factor, labour shortage, is more complex in its application to the 'homeland' situation. True agricultural production in a subsistence economy has varying labour demands, and labour for weeding or harvesting is in direct competition with labour demands in white agriculture, which competes for the seasonal labour of people in the 'homelands'. However, with unemployment rates close to 50 per cent, it is a rare village that has a real labour shortage. Rather, unemployed youth are unwilling to work for the very low returns from agriculture, especially if there are no cash wages. Indeed, anybody offering a cash wage, however low, can effectively employ as many labourers as are needed. Family and subsistence labour may be in short supply, because it is unpaid. Often land is underutilized not because of labour shortages, but because the resources needed to use land effectively are absent.

PAUL WEINBERG

ABOVE
Soil erosion in Transkei

Durning's fourth cause of land degradation, poverty, is clearly important. Family incomes in South African 'homelands' are among the lowest in Africa; migrant remittances, state pensions and local employment account for 90 per cent of village income. Agriculture, however, accounts for only 10 per cent. Malnutrition-related diseases, including kwashiorkor and pellagra in children and tuberculosis in adults, are widespread. Not only are impoverished families unable to invest in tree-planting or soil conservation schemes, but they are forced to strip the land of its resources. Families burn dung as fuel, instead of using it as manure; they strip trees for firewood, leaving hillsides bare. South Africa has few natural forests, and they are disappearing fast: of 250 distinct woodlands identified in KwaZulu fifty years ago, only one-fifth remain. In QwaQwa, forests have disappeared entirely. Women walk up to 10 kilometres a day to find wood, and expend far more effort on finding the fuel for cooking and heating than on growing it. Given the average per capita consumption of 200 to 800 kilograms per year, the fuelwood crisis is one of the most serious in South Africa.

State intervention: solving or exacerbating the problem?

The extent of soil erosion in any situation is influenced by government intervention, since the state can either assist a population to adapt its land-use system to deal with perceived problems, or worsen the pressures on the land, thereby intensifying erosion. Over the last four decades, the South African government has intervened more than almost any other in the name of conservation. Its land rehabilitation or 'betterment' programme was designed partly to address the massive problems of soil erosion caused by overcrowding. The programme's objectives were to enforce stock reduction to curtail overgrazing, and allocate common land for rotational grazing; to consolidate arable land and provide soil conservation measures; to draw scattered populations into closer settlements; and to provide nominal facilities and to improve soil conservation.

PAUL WEINBERG

Most betterment plans attempted to differentiate between 'full-time' and subsistence producers, thus creating or reinforcing inequalities within villages. Betterment depended on the concept of 'economic units'. As a result of land shortage, a typical betterment plan would create ten full units, twenty half-units and seventy one-tenth economic units. Allocating land according to these economic units meant that inequalities in a community were reinforced.

These programmes were fiercely disliked by communities who were unwilling to accept stock reduction, enforced differentiation and social engineering. Often betterment was forcibly imposed on communities, through manipulating tribal authorities or the use of the armed forces.

The merits of these betterment programmes are debatable, given the degree of social disruption they caused. Certainly, stock reduction was not enforced because of fierce community resistance to its imposition. In general, arable lands failed to maintain productivity. The construction of contours could have been achieved at far lower costs and without accompanying social dislocation, as was done in the sugar-growing areas of KwaZulu, or in the communal areas in Zimbabwe before independence. In some places, betterment was badly planned and hastily executed. The building of villages above arable lands intensified the erosion already taking place. In the arable lands themselves, contours were pegged but not constructed. Likewise, contours were built, but without waterways to lead the water safely away, thereby contributing to worsening erosion. The creation of villages meant that hills surrounding villages were stripped of wood.

All in all, enforced social engineering is an expensive, ineffective approach to soil conservation — yet the improving of land-use patterns has been largely ignored as an alternative by the government. Even though land use in the 'homelands' is based on a one-family one-plot system, this type of subsistence agriculture has received limited support in terms of research and technical assistance. Low-input agriculture and organic methods have been systematically ignored in favour of the technologies developed for white farms. On the whole, livestock farming in black rural areas has been dismissed as being traditional and backward.

Many of the assumptions made by government officials about stock farming in communal areas are rather dubious. One of these assumptions concerns stocking rates, and the carrying capacity of communal areas. Livestock systems which rely on intensive herding can be very efficient. A study of pre-independence cattle raising in southern Angola showed that pastoralists there were more efficient than white ranchers in Zimbabwe. In South Africa, the Richtersveld is an area where farmers have used ingenious methods to maintain yields for centuries even though conservationists consider that the nominal carrying capacity of the land has been exceeded. (See Chapter 8.) In Natal the veld of the Upper Tugela Location, a mere 100 000 hectares, has a better ecological composition than the neighbouring national park where the conservationists

LEFT
A village in Transkei perches on the edge of deep dongas formed by soil erosion

In Ciskei, 46 per cent of the land is moderately or severely eroded and 39 per cent of the veld overgrazed

of the Natal Parks Board keep people and animals out, and use fires to maintain the grassland. Livestock farming in the 'homelands' is not necessarily efficient — but the government's policies and practices have not improved matters.

A widening circle of erosion

The fuelwood and land crisis in the 'homelands' has grave implications for the environment throughout South Africa. For example, river systems are in danger: this is because 'homelands' tend to be located near major river catchment areas and when these are eroded, they dump heavy loads of silt into the rivers. This adds to the tendency of these rivers to flood during times of heavy rains, when the catchment areas are unable to absorb rainfall and release it slowly. The silt also dries the rivers up. Many of South Africa's major rivers run through or originate in 'homeland' areas. The Tugela, Caledon, Orange, Fish, Crocodile and Letaba Rivers have all been affected by erosion. The Upper Tugela Location, for example, has been earmarked by the KwaZulu Bureau for National Resources as a conservation area: they want the community to cooperate with conservationists by preventing their cattle from overgrazing the catchment. In the eastern Transvaal, the Letaba River is no longer perennial, partly because of increased population pressure in Gazankulu and its effect on catchment conditions. KaNgwane has seventeen new water extraction points on the Crocodile River; this adds to the pressure on water sources already taxed by irrigation, pesticide and fertilizer residues and industrial demand and waste.

Capitalist penetration in the 'homelands'

One of the state's only strategies for agricultural development in the 'homelands' has been to encourage commercial enterprises, whether in the form of smallholder schemes or plantations. The earlier schemes involved expropriating land from a community for a plantation. The owner would then employ the dispossessed people as wage labourers. Generally, these plantations exacerbated poverty, as people were worse off as landless wage labourers than as subsistence producers.

Ecologically speaking, these state-sponsored schemes not only affect the plantations, but also the land remaining in community hands. The plantations themselves, however, may be fairly well conserved. Tea, sugar, sisal, pine or eucalyptus plantations are usually designed with soil conservation in mind. Using pesticides or fertilizers may reduce soil fertility or biological diversity, but not necessarily more than a smallholder scheme would. On the other hand, pressure on remaining community lands leads to serious degradation. In Venda, half a community's land was confiscated for a tea plantation. Where the community was previously able to maintain a subsistence food economy, pressure on the remaining land now increased to the point where intercropping and crop rotation techniques were no longer able to sustain production. Rapid land degradation resulted.

Other types of commercialization also degrade land and increase erosion. Often land under a low-input system of production is subjected to a new high-input, monocropping system. Pesticides and fertilizers are used and intercropping techniques are replaced by monocropping. Returns to the farmers on these schemes are seldom better than those from subsistence agriculture. Sometimes, lands which are too eroded to be productive are nevertheless ploughed, or land close to river banks is ploughed, resulting in increased erosion. Such schemes also increase inequality within a community. Frequently the intention of development schemes is simply to create a class of better-off farmers. While the available technology benefits richer large farmers, the poorer farmers may be forced off the land onto even more marginal land, which then rapidly degenerates.

A new area of commercial interest in the 'homelands' is forestry. This has significant environmental implications, owing to the effects on arable land and grazing. Moreover, within a catchment area forestry uses water excessively, and may affect not only the water table but also water availability to other crops. At present, most commercial forestry in the 'homelands' is limited to state plantations. But in KwaZulu, for example, there are already 750 hectares of commercial timber, and a further 2 000 hectares are planned. On a larger scale, especially if trees are planted in catchment areas, forestry could have a major impact on an environment already under severe pressure from population growth. (See Chapter 5.)

The impact of soil erosion in the 'homelands' is much more noticeable than in the commercial farm sector where white farmers control the land. However, this sector is not entirely healthy. A deepening debt crisis, significant inequality in the sector, and severe soil degradation, have been a few of the problems.

South Africa's agriculture has a fairly poor resource base. Only 11 per cent of the land is arable, and less than 3 per cent is regarded as highly productive. While the climate is erratic, there are generalized ten-year wet and dry cycles. During the wet cycles, erosion levels increase as farmers tend to plough up marginal land which then erodes in dry cycles; they also tend to increase their livestock population densities during wet cycles, but do not reduce stock numbers sufficiently in dry cycles (Huntley et al, 1990:47). More than half the country receives less than 500 millimetres of rainfall per annum, the minimum amount considered necessary for crop farming. Moreover, 20 per cent more land is ploughed in the marginal, western half of the country than has arable potential, while in the eastern half of the country, more than 20 per cent of land with arable potential is not utilized.

Overcrowding is undoubtedly the major cause of soil erosion in South Africa

There is increasing differentiation among white farmers. One-third of them, about 17 000 farmers, are responsible for 80 per cent of all the produce. The poorer farmers are more likely to abuse the environment by overstocking their farms or trying to plough where they can. Richer farmers use more pesticides, fertilizers and mechanization and are more likely to monocrop the land. On the other hand, in livestock areas they are more likely to maintain sustainable stocking rates, and to plant fodder crops for their livestock.

The more arid areas of the country particularly need careful management. The semidesert Karoo advances steadily north-east, faster in dry years and more slowly in wet. In the north-western parts of the country, where semidesert conditions exist, overgrazing results in serious land degradation. Government policy encourages stock reduction based on scientifically-determined stocking rates, and through subsidy, encourages the planting of permanent pastures. Yet it seems that this policy has not managed to slow down the advance of the Karoo.

The use of pesticides

Currently, South African farmers are totally dependent on pesticide use for successful crop and livestock yields. The use of pesticides has grown dramatically since the 1950s, especially in crop farming. Most crops are sprayed or treated with at least three different kinds of pesticides. For example, maize may be treated with herbicides to control broad-leafed weeds and grass, and with insecticides to eliminate cutworm and maize stalk-borer. Fresh vegetables like lettuce may be sprayed up to ten times, with as many as four different pesticides. Farmers are caught up on a pesticide treadmill. The more they use, the more they need.

Sometimes, pesticide used by one group of farmers to solve one problem causes even greater problems to other farmers. The experience of farmers in the Tala Valley in Natal, themselves heavy users of pesticides, reveals one of the many disadvantages of this approach. Their vegetable crops are being damaged by spray drift of hormonal herbicides from sugar cane fields, causing damage valued at millions of rands annually. Mark Laing, a plant biologist at the University of Natal, says: 'Government researchers have found that 33 to 100 per cent of rainfall samples contain "hormone herbicide"; on the fresh produce markets, it has been difficult to find tomatoes or lettuce on sale which do not display typical, textbook symptoms of hormonal herbicide damage' (*Indicator*, 1990:41). It is estimated that up to 20 per cent of spray from a tractor, and 80 per cent from an aircraft, drifts off target; 15 per cent of that remains permanently dispersed in the air.

The Tala Valley farmers took the pesticide manufacturers to court; but their claim was rejected on the grounds that the user, not the manufacturer, was responsible. The Department of Agriculture has agreed to ban the use of 2,4-D in the Tala Valley, pending investigation. But, according to Laing, a local ban is nonsensical. The possible solutions are to ban the use of hormonal herbicides altogether, or to set strict rules governing the use of herbicides. At present, laws restricting pesticide use are practically non-existent.

One-third of our 50 000 white farmers are responsible for 80 per cent of all produce

Pesticide use on farms causes as many problems as it solves. Pesticides eliminate not only pests, but also beneficial insects which help to control the harmful ones: farmers are then forced to use more pesticides to combat the escalating problems. Pesticides

ABOVE

Crop-spraying in the eastern Transvaal ... richer farmers use more pesticides

are also detrimental to the environment and human health, some even having carcinogenic and mutagenic effects. Pesticide residues are found in food, water supplies and in the air. For example, high concentrations of 2,4-D have been found in silt at the mouth of the Umgcweni River in Natal, leached from sugarcane fields; and 2,4,5-T is thought to be linked to birth defects among forestry workers in Natal. Both these hormonal herbicides were ingredients of the notorious Agent Orange defoliant used during the Vietnam war.

Pesticides banned or restricted for use in their country of manufacture, including 2,4-D, are widely used in South Africa. The government is one of the worst offenders. The Department of Environment has used 2,4,5-T to control weeds on plantations, and has used BHC to control locusts in the Karoo. Both these pesticides are among the 'Dirty Dozen', a list of the twelve most noxious pesticides which the Pesticide Action Network believes should be banned internationally. Furthermore, six of the 'dirty dozen' are widely used in South Africa, in agriculture and by public authorities. One of these, DDT, is used only for mosquito control, which is approved by the World Health Organization. Recently, the South African Police announced they would use paraquat to destroy cannabis (dagga) plants. (See Chapter 2).

Workers who use pesticides are generally untrained and are not often given the necessary protective clothing. Labels detailing appropriate spray precautions are seldom understood by illiterate workers. Those who come into contact with these poisons may suffer acute poisoning, or longer-term chronic effects. Many deaths or injuries from pesticide use result. The ILO estimates that 1 600 South Africans die from pesticide-related events each year (*Critical Health*, 1990:76).

Only 11 per cent of South Africa's land is arable, and less than 3 per cent is regarded as highly productive

The pesticide industry, which is controlled by two dozen multinationals worldwide, invests heavily in research and development for new chemicals. By dominating research, these companies have considerable influence over the technology farmers use. Although modern pesticides are more sensitive to problems of environmental persistence, they do nothing to promote farming systems less dependent on pesticides. Integrated pest management (IPM) is a safer, more environmentally sound and more viable approach to controlling pests. This is a system that combines natural methods, cultural practices, careful field inspections and the cautious use of pesticides. It is currently being used in the cotton and citrus agricultural sectors. In Nicaragua, IPM reduced pesticide use in cotton by 45 per cent in the 1970s.

Forestry

One of the major structural changes taking place in agriculture is in forestry development. In South Africa more than 1,2 million hectares of land is afforested (out of 85 million hectares in total), and this area is growing by 30 000 hectares per year. Over 50 per cent of timber is used for pulp and paper and a lucrative export market has developed.

Forests created by humans, or plantations, have an enormous impact on the environment. They have earned the title 'green deserts' because, unlike natural forests, they do not encourage biodiversity. Trees in a plantation have a high water demand, and to obtain a satisfactory return most plantations have been established in the escarpment areas of the country, which are also the water catchments of our major rivers. So forestry replaces natural grassland, which allows surface water to run off and be used lower down for irrigation or human consumption, and replaces it with trees which absorb some or all of the water previously entering the river system. For the company growing the trees and making the pulp, this is profitable; but for the downstream user, the effects can be devastating. Communities in KwaZulu, for example, are experiencing water shortages as springs dry up ; rainfed agriculture has become marginal as the water table has dropped because of the demand by plantations.

Planted forests have other long-term detrimental effects. Steeply sloping land is sometimes ploughed to boost production, adding enormous silt loads to rivers and hastening soil loss. Forests also necessitate heavy use of pesticides; these chemicals can drift into the air or dissolve in rainwater and affect communities downstream. Plantations also compete with other forms of land use. To make space for forests, farming is often pushed onto more marginal land, thus intensifying soil erosion.

Plantations have earned the title 'green deserts' because, unlike natural forests, they do not encourage biodiversity

The economic implications of corporate takeover of massive areas of land for forestry needs close examination. Questions need to be asked — such as, is the Department of the Environment serving South Africa's interests by issuing quotas which allow Mondi and Sappi, the two forestry conglomerates, to buy up huge tracts of grassland and convert them into forest? While a democratic government may be able to represent all South Africans in these decisions, the present government clearly does not necessarily take into account the effects on the poor, disenfranchised communities who may be most directly affected by such environmental restructuring.

Is the Department of the Environment serving South Africa's interests by allowing Sappi and Mondi to convert huge tracts of grassland into forest?

Land reform and the environment

It is widely recognized that land redistribution is an essential component of the creation of a more democratic South Africa. South Africa's land ownership ranks among the most inequitable in the world: whites control 87 per cent of the land while blacks hold only 13 per cent. The 50 000 white farmers have access to twelve times as much land for cultivation and grazing as the 14 million rural blacks.

This unequal land distribution in South Africa contributes to environmental degradation. Overcrowding in the 'homelands' and the negative patterns of large-scale white agriculture are opposite sides of the same coin. Will land redistribution, however, reduce soil erosion? Before this question can be answered, it should be noted that land redistribution is enormously difficult to accomplish on a massive scale. According to Stan Sangweni of the ANC

> Since 1980, resolutions have called for governments and others responsible to put in place appropriate land redistribution measures ... A recent assessment indicated that very little has been put in place as a result of these resolutions. Governments clearly don't have the political will to implement the resolutions. One obvious reason is that the owners of huge tracts of land are in the government themselves. In Africa, most countries fighting for their independence in the 1950s and 1960s promised that their first task would be to redistribute the land, but I cannot think of any country where it has actually taken place. On the contrary, in practically all African countries, the land arrangements are virtually the same, except there's been a change in ownership' (*New Ground*, 1990:2).

The debate surrounding land reform in South Africa has only just begun. Both the Urban Foundation and the DBSA have called for limited land redistribution. Because they have made concrete proposals it is worthwhile to examine these in some detail to analyse their potential environmental implications. According to the DBSA, it is possible to redistribute up to 8 million hectares of land which is either underutilized or heavily indebted, without significantly affecting current production. A total of 2 million hectares is trust land, due to be incorporated into the 'homelands'. About 80 per cent of these 2 million hectares are in northern Natal and the eastern Transvaal. Land in the remaining categories includes farms owned by indebted white farmers, smallholdings around cities and church land. The DBSA says, 'We have to get the fictitious assumption that smaller farms would mean less efficient production out of the way.' According to the Urban Foundation, the entry of black farmers 'would lead to some farms being farmed worse and

A study of cattle raising in pre-independence southern Angola showed that pastoralists there were more efficient than white cattle farmers in Zimbabwe

some better ... if 10 per cent of the current white-farmed land taken out all over the country disappeared from farming entirely, it would have no significance as far as availability of sufficient food is concerned ... in many parts of the country, white-operated farms are producing only a percentage of what could be produced under top management' (Dolny, 1990).

Unlike the Urban Foundation, the DBSA is opposed to resettlement schemes, which it views as too costly and ineffective. The Urban Foundation's redistribution proposal is based on state land purchases and redistribution in settlement schemes. However, both the DBSA and the Urban Foundation see the market system (the buying of land) as the only effective mechanism for redistributing land. For redistribution to happen at all, both call for subsidization to enable black farmers to purchase land. They also underline the need for credit, extension and market support. The DBSA estimates that it will cost the state only between R1 and R2 billion to

redistribute the 8 million hectares. Others put the estimate at close to R20 billion if market mechanisms are to be used. The DBSA says the 8 million hectares can be redistributed to 500 000 farmers — an average of 16 hectares per farmer. This would result in the creation of a massive number of smallholdings throughout the country, especially in the wetter eastern half. Even assuming that proper support could be provided for these farmers, including credit, would this really assist in making agriculture more sustainable?

In the ecological zones suitable for crop farming, it is likely that redistribution would boost production. More labour-intensive cultivation would result in small growers outperforming large farms, as is the case with sugar growers in Natal. Land which is at present underutilized could be used productively. But if the methods of production are modelled after those of most white farmers, it is doubtful that agriculture on that land would be more sustainable or economically viable. Secondly, the type of land suggested for redistribution, especially from indebted farmers, may not be of the quality needed for arable farming. Thus farmers might well grow crops in areas not ecologically suitable, out of a need to increase incomes.

In fact, one of the most serious environmental issues raised by redistribution policies is the question of land quality. All too frequently, as in the case of the DBSA proposal, redistribution tries to target the land which is available. If this is marginal or fragile land, then redistribution to smallholders can be disastrous.

ANC researchers on the land question have argued that for redistribution to be effective, productive land (most of which is held by productive white farmers) must be redistributed as new farmers will be unable to sustain environmentally sound agriculture without adequate resources. At the very least, redistribution must concentrate on the areas where the land has the potential to be productive.

It is, however, very unlikely that large quantities of productive farmland can be made available for redistribution. The cost is likely to be prohibitive and the effects on food and raw material supply will deter attempts at confiscation. Thus land reform is more likely to rely on poor quality land or land which has potential, but is undeveloped, unless politicians can be persuaded to overcome short-term practical obstacles in favour of more long-term considerations of viability.

BELOW
Shepherds look down on Sunduza village, a betterment village in Transkei

LESLEY LAWSON

Restructuring through rural development

Irrespective of the complexities of redistribution, there is an urgent need to restructure agriculture. Land in the dry western parts of the country needs to be taken out of arable production, while in the wetter eastern parts, more land could be ploughed. In the semi-arid livestock areas, herd sizes need to be limited to the carrying capacity of the grazing in dry years. In bushveld areas, substitution of domestic animals for game farming is likely to improve long-term economic viability.

Even in the 'homelands', effective sustainable development

'We have to get the fictitious assumption that smaller farms would mean less efficient production out of the way'

— Development Bank of Southern Africa

could improve people's living standards. The Zimbabwean experience shows that building the communal areas' economy is an effective development strategy. At the very least, the potential of these areas to support their population needs to be used as part of a broader land reform strategy. This will involve environmental reconstruction. Earthworks alone will not suffice; reconstruction needs to be holistic. It needs to begin with the needs and interests of people, and their active participation is a minimum condition for success. Within communities, a range of different interests is involved: reconstruction depends on improving land use, which means that landholders and cattle owners must be involved. Projects must also benefit the landless, since their active participation is essential.

A broader rural development strategy is more likely to be successful than attempts to boost agricultural production. For example, rural development should attempt to meet the water and fuel needs of a community. This may involve electrification or other sources of energy.

Can agriculture become sustainable?

Irrespective of the exact outcome of land reform, there is a need to start thinking seriously about sustainable agriculture — a need to find a way of farming that will ensure improved land and water resources over time.

Present farming methods are reducing soil fertility and depleting water resources for future generations of South Africans. The alternative is sustainable agriculture, which is based on the idea that farming should proceed in harmony with nature, using natural resources rather than exploiting them, and promoting natural conditions where possible.

Some ways in which sustainable systems could be applied to agricultural problems are suggested below.

Preventing soil erosion

Soil erosion occurs when soil is left exposed to wind and rain; thus the vast areas of ploughed land on South African farms cause soil erosion. The sustainable alternative is to keep soil cover by using mulch and minimum tillage methods, and by planting grass banks between ploughed lands.

Protecting soil fertility

Soil fertility need not (and should not) depend on chemical fertilizers. Soil is a living thing, and its fertility depends on its ability to hold water and plant foods. This in turn depends on a rich organic life in the soil. Organic farming makes use of manure and compost to regenerate the soil, but in South Africa's hot, dry conditions there is not enough compost and manure for field crops. There are, however, other forms of soil enrichment:

- Grass leys rest the soil and improve its structure.
- Crop rotation (planting different crops in the same field in different years) allows a crop which takes nitrogen out of the soil to

be rotated with a legume crop which replenishes nitrogen.

- Intercropping involves growing two crops side by side so that each assists the other's growth.
- In agroforestry, crops and trees are planted together. Leguminous (nitrogen-replacing) trees can put plant food back into the soil. Trees also help to draw water and food to the surface from deep within the soil, which obviously assists the crops.

Avoiding harmful pesticides

In the area of pest control, sustainable agriculture uses nature's abilities to fight pests.

- Crop rotation can help to break the life-cycle of weeds and insects.
- Planting can be timed carefully to avoid damage by pests.
- Natural enemies of pests can sometimes be introduced (judiciously).
- A small amount of insect damage can be allowed, if the farmer works out the costs carefully. The loss of a small part of a crop can be less expensive than the costs (financial and otherwise) of pesticides.

Farming with livestock

Livestock is particularly important in sustainable agriculture. Animals are an important source of manure, which is vital to soil fertility. Fodder can be grown for animals in a sustainable way, by means of planting grass pastures. These are less expensive than other forms of animal food, such as grain. When using natural pastures, the farmer has to limit the number of animals on the land to prevent overgrazing. If this is done properly, the use of chemical additives, hormones and pesticides can be significantly reduced.

Effecting change

Some white farmers have begun to see the need for different methods of agriculture, mainly because of the rising costs of present methods. Many have cut down their use of fertilizer. Some are questioning whether pesticides really solve their problems. Over the last ten years prices of chemical inputs have increased much more than prices of agricultural products, causing debt for many farmers. The methods of the 'Green Revolution' — high chemical input farming — developed in the 1960s, claimed much higher productivity than other methods. But this claim has proved true only in good years. In dry years, yields are just as low as when no expensive inputs are used. It is only the multinational companies who manufacture the pesticides, fertilizers and improved seed, whose profits have grown consistently.

Part of the reason why sustainable agriculture is not in widespread use at present is that the farmers' support services are dominated by the multinational companies who make the chemical inputs. If farmers are to change their way of farming, they need support from the industries servicing their sector and from the

Having sustainable agriculture as a farming policy will mean that more small farmers, many of them poor rural people, can make a living and a contribution to the country's agriculture

state. If the service industries are profiting from the present system, is there any motivation for change?

The uncertainty and risks of change have also contributed to the resistance to sustainable agriculture. Farmers using the biological methods will need expert advice on the timing of planting, cultivating, watering and pest control. Sustainable methods might imply new crops and new machinery, or require better trained labour. Farmers are unlikely to make these changes if there is no research evidence that these methods will be successful.

Farmers need economic incentives too. What is required is not only a change in farming methods, but a change in the economic policies which govern agriculture. Present research and support services are geared to large-scale single crop farming. Tax incentives and pricing policies favour large farmers, not small ones.

It is a political necessity to rethink agriculture to make it more sustainable. Much needs to be done to convert unprofitable cropland to pastures, to cut down animal numbers in many areas to match grazing capacity, and to use high potential land more effectively, especially land in the 'homelands'. Appropriate crops need to be investigated — for example, is sugarcane the best crop for the highly productive Natal coast? Overall, attention must be paid to environmental conservation and the prevention of soil erosion.

Sustainable agricultural methods are also politically desirable because they will create more employment. They are better suited to poorer farmers, since they are labour-intensive and require low cash inputs. Introduction of sustainable methods should therefore be part and parcel of land reform as we move towards a more democratic society. Having sustainable agriculture as a farming policy will mean that more small farmers, many of them poor rural people, can make a living and a contribution to the country's agriculture.

However, farmers lacking in resources will not automatically take up the methods of sustainable agriculture without incentives and training. The present model of 'success' will lead them to try to imitate the methods of the richer, large-scale farmers of today. To gain widespread acceptance of sustainable methods, state subsidies need to be redirected, new research must be undertaken and extension workers retrained. Only this will result in the popularization of sustainable agricultural systems.

13

THE OCEANS

Our common heritage

◆ FRANCIS MANUEL ◆ JAN GLAZEWSKI ◆

The Law of the Sea Convention compels coastal states to protect and preserve the marine environment from over-exploitation and pollution

The oceans of the world: rights and duties

The world's oceans, covering over two-thirds of the earth's surface, have been enjoyed, navigated, explored and fished by people since the dawn of history. It was not until the fifteenth century though, when sea-faring nations began to compete for resources, that the first claims were made to sovereignty over parts of the sea. By the seventeenth century opposition to such claims had grown considerably, and since then, international law governing the use of the oceans has come to accept the idea that the sea is an area of common heritage, with the exclusion of narrow bands of 'Territorial Waters' adjoining coastal states. This principle has been reinforced by the establishment of 'high seas freedoms' — freedom of fishing, freedom of navigation, freedom of overflight, and freedom to conduct scientific research. More recently, technological developments have led to the recognition of further freedoms, for example, the freedom to lay pipelines and underwater cables.

The inclusion of the exploitation of the sea's resources among these freedoms was based on the traditional belief that they were inexhaustible. However, the dramatic escalation in fishing effort and technological development after World War II led to a spectacular drop in the yields of some traditional fishing grounds, including those of South Africa. This resulted in a realization of the need to limit access to fisheries resources, and to provide equitable means of doing so. A similar assumption that the capacity of the oceans to assimilate waste was infinite, was dispelled by incidents such as the outbreak of Minamata disease in Japan in 1953, when forty-three people died, and sixty-eight were paralysed after eating fish contaminated with mercury. In the last two decades especially, there have been a number of attempts to control marine pollution at an international level.

Increased control over fisheries and marine pollution has been achieved in part by granting coastal states greater control over adjacent seas, even though this has detracted from the principle of the freedom of the high seas. Initially such control was limited to Territorial Waters of 3 nautical miles. During the 1960s and 1970s, however, there was considerable pressure from coastal states to extend, in particular, the fishing zones. Such claims have been validated in the 1982 Law of the Sea Convention (LOSC), which establishes the right of coastal states to an Exclusive Economic Zone (EEZ) of 200 nautical miles.

Although this Convention has yet to come into force, many of its provisions are already being observed. The LOSC also allows coastal states to claim Territorial Waters of 12 nautical miles, a contiguous zone of 24 nautical miles, and rights to the resources of the continental shelf. At the same time, the LOSC imposes duties on the coastal states, among others the duty to protect and preserve the marine environment from both over-exploitation and pollution. Since 90 per cent of the world's fish stocks occur within 200 nautical miles of the coast, and since 85 per cent of marine pollution comes from land-based sources, these provisions could have a significant impact. Moreover, the LOSC imposes both rights and duties on all states with regard to the protection and preservation of the high seas.

BELOW

Heavily-laden pelagic fishing boats return to harbour in False Bay with the day's catch

The status of South Africa's coastal environment and resources

In 1977, South Africa extended its jurisdiction over the living marine resources of its coastal waters by declaring an exclusive fishing zone of 200 nautical miles. South Africa also claims Territorial Waters of 12 nautical miles. As outlined above, in terms of the LOSC — which South Africa has signed but not yet ratified — such claims go hand in hand with responsibilities, both to the international and the regional community. The question is, to what extent are we fulfilling these responsibilities?

A second question concerns the belief that the sea and its resources lie in the public domain. This philosophy is not restricted to the international sphere, but is also reflected in the domestic laws of many countries, including South Africa. Roman law, which provides the underlying foundation of South African law, classified the sea and seashore as *res omnes communes*, which meant that they were subject to the enjoyment of all, and not available to individual ownership. This classification has been modified in South African law, but the philosophy is inherent in the relevant legislation. Similarly, fish and other marine resources are regarded as *res nullius*, which means they belong to nobody until such time as they are caught, when they become the property of whoever caught them. To what extent has this traditional regime been altered by present legislation?

Uses and abuses of the coastal zone

South Africa is blessed with over 3 000 kilometres of coastline, and is well known for its beautiful beaches. These beaches and other coastal features form an integral part of the marine environment, and any discussion of the marine ecosystem and human influence on it, would be incomplete without reference to them. The coastal zone — including the near-shore waters, the intertidal zone, and a variable area landward of that — is a highly sensitive region in which the land, the oceans and the atmosphere interact in a unique way. It incorporates geographic features such as coastal lakes, estuaries and dunes, all of which are easily disrupted by inappropriate development. At the same time, the coastal zone is also the most densely populated part of most coastal countries. In the United States, for example, one-third of the people live in the 15 per cent of the land area that constitutes the coastal zone; in the Cape, 90 per cent of the population live within 100 kilometres of the coast. There is, therefore, obviously a need for strict control over human activities in this area.

The intertidal zone has long been recognized by South African law as not being subject to private ownership, which in itself should provide some measure of protection. The Seashore Act of 1935, which is one of the chief contemporary acts regulating the area, vested ownership of the sea and the seashore in the Governor-General. With the establishment of the Republic in 1961, owner-

THE ARGUS

In the Cape 90 per cent of the population live within 100 kilometres of the coast

The Reservation of Separate Amenities Act of 1953 excludes 'non-whites' from those beaches with the best-developed facilities

ship passed to the State President. Both these officials were figureheads and were in effect trustees of the area for the people, rather than owners in the traditional private capacity. Today we have an executive State President who presumably still retains this custodian's role.

An attempt was also made to control developments in the broader coastal zone some years back, when regulations pertaining to proposed developments within a 1 000 metre strip extending from high-water mark were promulgated under the Environment Conservation Act (1982). This Act has since been repealed, and the status of these regulations is unclear. Unfortunately, the law has also not clearly defined the custodianship role vested in the State President, and as a result there have been many unsatisfactory developments in this area. Of the examples given below, some have made inroads into the rights of the public, while others have led to degradation of the environment.

Beach apartheid

A prime example of the erosion of public rights in the coastal zone was the designation of different beaches for use by different race groups. Such designation was effected in terms of the Seashore Act — which specifically provided that any regulation could differentiate between various classes or kinds of users of the seashore or sea — and the Reservation of Separate Amenities Act of 1953. In effect they resulted in the exclusion of 'non-whites' from those beaches with the best-developed facilities.

The validity of such legislation was challenged as far back as 1943 in the case of Rex versus Carelse. The accused in this case, a

CSIR

RIGHT
The effects of building on a primary sand dune. The sand dune is establishing itself and encroaching on houses built on the beach at Fish Hoek, Cape

'non-European', was found guilty of bathing at a spot on the Strand which was set aside for 'Europeans'. One of his grounds of appeal was that the regulations were invalid because they militated against the common law rights of the public to swim in the sea. He argued that regulations should be confined to general matters such as the wearing of bathing costumes or the demarcation of dangerous areas. The court dismissed his argument and ruled that the regulations were in conformity with the ambit of the Act. In adopting this interpretation, the court virtually buried any hope of maintaining the common law principle of public access to the sea and seashore.

More recently, these laws were challenged by the Mass Democratic Movement in the form of beach picnics, and by Allan Hendrickse, Leader of the majority party in the House of Representatives, in his (in)famous swim in Port Elizabeth in 1987. The Separate Amenities Act was finally repealed in 1990 and the Seashore Act amended to do away with this form of discrimination.

Some local authorities have allowed development to encroach right over the dune system, with disastrous consequences

Coastal developments

The Seashore Act regulates in some detail which type of developments may be allowed below high-water mark. Those which it does allow are all of a public nature, for example, restaurants, bathing boxes, jetties, and harbours. Exceptions can be made in cases where the shore was alienated to private individuals prior to the passing of the Seashore Act, or by special resolution of Parliament. Recently, though, there have been attempts to extend the list to include private developments, and even suggestions that the Seashore Act be amended. Moreover, the Cape Provincial Administration — which has been responsible for screening devel-

BELOW
The destructive effect of waves associated with spring high tides at Fish Hoek

opment proposals in the province since 1986 — appears to be ignoring these provisos, and has approved several private developments which involve the alienation of parts of the seashore. In some cases, such as the proposal to fill in and develop an area of the sea adjacent to the Robberg Peninsula near Plettenberg Bay, vigorous public opposition has caused the project to be shelved. On the other hand, the Club Mykonos development in Langebaan has been allowed to go ahead, and has the time-share project at the Strand.

Developments such as these in the coastal zone have resulted in the degradation of the environment and in particular, of features such as dunes and estuaries.

Dunes

The dune system forms a highly dynamic transitional zone between the land and the sea. On the seaward side, the dunes are transient and mobile, and function as reservoirs of sand during periods of erosion. More established dunes at the back of the beach are stabilized by vegetation, and provide protection for the areas behind them. The system is essential for the stability of the beach, and interference in its functioning can have disastrous and costly consequences.

In spite of this, some local authorities have allowed development to encroach right over the dune system. Apart from destabilizing these areas, buildings standing in the path of dunes are subject to extremely high maintenance costs — as the owners of beach-front houses in Fish Hoek know only too well.

Similarly, many roads or footpaths have been built over dunes which are normally eroded during winter storms. On the other hand, attempts at artificially stabilizing dunes have also had unforeseen consequences. For example, the stabilization of dunes in the Cape Recife area has led to the erosion of King's Beach, one of Port Elizabeth's main bathing beaches, because the natural sand supply has been cut off. A costly engineering solution has had to be implemented to remedy the problem.

Dunes have also been affected by mining projects. For instance, the exploitation of precious stones — particularly diamonds — on the West Coast has resulted in the relocation of tons of beach sands. Proposals to mine heavy mineral sands in the St Lucia area in Natal, and in areas on the West Coast, are under serious consideration. Historically, mining law has developed around the needs and interests of inland industry, and is inadequate when applied to the problems associated with mining in coastal areas. Nevertheless, the draft Minerals Bill, due to come before Parliament in 1991, makes no special provision for the coastal zone.

Virtually all of the 365 estuaries around our coastline have been affected by human activities

Estuaries

Estuaries are another sensitive and vital component of the coastal zone. They are highly productive and serve as breeding and nursery grounds for a variety of marine species. They are also important habitats for large numbers of resident and migratory birds. They are, however, particularly vulnerable to human activities. Their natural beauty and the shelter they provide make them natural harbours as well as popular sites for both residential and recreational facilities. They are also exposed to the impacts of activities in their catchments, such as the abstraction of water, bad agricultural practices, the discharge of industrial or municipal wastes, road construction, and floodplain development. Their survival is therefore dependent not only on coastal zone legislation, but also on the adoption of more holistic water and agricultural policies.

Mtwalume estuary in the 1930s

A THORPE

Mtwalume estuary today

GEORGE BEGG

ABOVE

The Mtwalume estuary was typical of Natal's unspoilt estuaries in the 1930s, but today the estuary is badly silted as a result of construction

Virtually all of the 365 estuaries around our coastline have been altered or affected by human activities to some extent.

The Mtwalume estuary on Natal's South Coast is a good example. In the 1930s the Mtwalume estuary was up to 6 metres deep, and extended 4 kilometres inland. The mouth at the southern end was relatively stable, and the estuarine life abundant, including species of fish never seen in estuaries today. Conditions changed after the expansion of sugar farms to the banks of the river and ultimately onto the floodplain. The natural vegetation holding the banks was destroyed, opening the way for massive siltation. Then embankments were built across the floodplain to support a new railway line, thus restricting floodwaters and causing the deposition of the silt load in the estuary. Today the estuary is reduced to a lagoon some 700 metres long, and no deeper than 1,5 metres. The mouth is highly mobile, and the waters are regarded as barren.

An initial step which could be taken towards rectifying this unsatisfactory state of affairs in the coastal zone, would be for the Minister of Environment Affairs to take the initiative and declare the sensitive coastal zone a 'limited development area' — which he is entitled to do under the current Environment Conservation Act.

Fisheries and bene(fish)iaries: our common wealth or our common burden?

South Africa is fortunate in that the oceanographic conditions off its coast create a very favourable habitat for a number of marine

TOO MANY RULES AND REGULATIONS

REMINISCING with some of the older residents of White City and Diazville — the 'coloured' townships of Saldanha Bay — gives insight into the bitterness underlying their stated opposition to rules and regulations. Many of them grew up in Saldanha Bay, living on or close to the beach-front, having 'mussel picnics' with the whole family, and, very often, making their entire livelihood from fishing. Most families had boats or dinghies and fished with trek-nets in the bay, selling their catch privately or making it into 'bokkoms' by drying it in their front yards. They were people of the sea.

The first changes came with the Group Areas Act. People were forced to move away from the seafront, some lost their businesses, and families were split up. They were restricted to four small beaches close to the fishing harbour, which were later filled in to build lobster tanks. Then the Department of Sea Fisheries started introducing regulations. Boats had to be licensed, and to be licensed they had to be equipped with motors, which many could not afford. Most of the permits for trek-netting were, therefore, acquired by whites, who were more easily able to finance their boats and equipment. Restrictions were placed on the private sale of fish and lobsters, and on the collection of other species, even for private use. Then the municipality prohibited the drying of fish on residential property. After 1976, with the development of the harbour, even further restrictions were placed on the movement of boats within the bay.

In the end, most of the fishing community went to work for the fishing companies. But, although they appear to live relatively comfortably, the work is seasonal; in some cases fishermen get paid according to their catch; and there is little job security. The Food and Allied Workers' Union (FAWU) reports that in the last year over 500 workers have been retrenched from jobs in the fishing industry in the western Cape, while 2 500 others have been on short-time since 1985. Most of all, though, they have had to adapt to a completely different lifestyle, brought about by laws in which they have had no say. Small wonder that most of their children have become alienated from the sea. ❐

Once a common resource is being exploited by more than one harvester, there is little incentive for restraint by competitors

species. These resources are highly productive, and the wholesale value of marine species which are commercially exploited runs in excess of R1 to R1,5 billion each year. There is huge capital investment in the industry, and substantial employment is created, either through direct fishing activities or through related activities such as boat-building and maintenance. The major resources along our coastline are the offshore fish species — hake, sole, anchovy, pilchard, horse mackerel and round herring — inshore resources such as the West Coast rock lobster, and rock lobster, perlemoen and squid on the South Coast. In addition, a number of linefish species are caught by both commercial and recreational fishermen, using a mixture of traditional and modern fishing techniques. Recreational interest in linefishing is considerable, to the extent that the annual investment therein is thought to exceed the annual wholesale value of all South African marine resources together. This sector should therefore not be underestimated as a source of job creation and political pressure. A number of the resources mentioned above, such as the rock lobster, are the basis of valuable export industries, conferring additional economic and political importance with regard to the balance of trade. In short, there is a clear imperative to ensure sustainability in the development and utilization of South Africa's marine resources.

These marine resources belong to us, the people of South Africa, and are managed and exploited on our behalf by the Minister and the fishing companies respectively. It is, therefore, our right to know just how well this is being done, and whether, in fact, catches are being kept to sustainable levels. Some of the historical evidence suggests that this has not always been the case. Before elaborating on this, however, it is necessary to look at some of the problems associated with fisheries management. It is our intention that in dealing with these issues, we shall not be supplying solutions, but widening the policy debate to include the whole community.

The argument was, 'If we don't catch them, they will'

The tragedy of the commons

The basic problem of any fishery is precisely its common nature, with the inherent implication that it should be accessible to anyone. Up until the post-war period, when worldwide fishing effort increased dramatically, the favoured management approach was indeed one of open access, based on the assumption that market forces of supply and demand would suffice to regulate fish stocks. The series of spectacular crashes in many of the important world fisheries during the 1950s and 1960s showed, however, that human demand for fish had far outstripped the stock replenishment rates. The most dramatic crash was perhaps in the Peruvian anchovy fishing industry, which dropped from 13 million tons in 1969, to 4 million in 1971, and to less than 1 million in 1976.

This phenomenon is a good illustration of the 'tragedy of the commons' and suggests that over-exploitation is the inevitable outcome of mixing a common property resource and free market competition. Simply put, once a common resource is being exploited by more than one harvester, there is little incentive for restraint by competitors. The resource not taken by the cautious harvester will simply be taken by others. Alternative methods of management have therefore had to be formulated.

In South Africa, as in many other countries, the approach has been a combination of regulations including limited access based on quota allocations, the establishment of a total allowable catch, season closures and size restrictions. In spite of both this and dramatically improved management practices for the major species since the early 1980s, problems have persisted, and there are indications that some of our stocks may be over-exploited. The example discussed below points to some of the possible reasons behind this.

The anatomy of a crash: Namibia

As in South Africa, in the 1950s and 1960s fisheries in Namibia were managed with a mixture of general ad hoc strategies, of which annual quota setting was the most significant. During this period, pilchard quotas were kept at a moderate level, and thus maintained a stable catch of 200 000 tons per annum. Then, from 1963, improving technology, greater capitalization, and increasing pressure from South African companies began to push up quota allocations. As quotas were increased — against scientific advice — the exploitation shifted from the yearly surplus stock to the main parent stock. In other words, it began to erode the actual 'capital'. In 1967 the

THE GILL-NET FACTOR

THE use of gill-nets is widely perceived to be contributing to a reduction in fish catches.

The term gill-net is, however, a wide one, and includes anything from nets used by traditional coastal fishermen to the shark-nets used in Natal and the large-scale pelagic drift-nets commonly used by Japanese, Taiwanese and Korean fishing boats. It is these latter nets, which can be up to 65 kilometres long and 15 metres deep, which are the main cause of concern, although it should be pointed out that the shark-nets in Natal have also had a significant impact on the coastal ecology.

Drift-nets are a particularly wasteful and indiscriminate form of fishing, ensnaring a high percentage of non-target species, including dolphins, small whales, seals and turtles. Their efficiency can also lead to over-exploitation of target species, which in our waters would primarily be tuna. South Africa does not, however, issue permits for the use of such nets in local waters. Furthermore, after reports in April 1989 that certain vessels were employing them, regulations were issued prohibiting these nets from even being carried on boats entering our waters. New regulations published in July 1990 also prohibit transshipment within our ports of tuna caught with drift-nets. In spite of this the controversy continues: there are allegations of continued illegal use of drift-nets; entry permits for Cape Town harbour were issued to a number of vessels carrying drift-nets; and there is no legislation preventing the transshipment at sea of fish caught with drift-nets, or the re-supply of vessels employing them, in or from our ports.

While the tendency for the local fishing industry to lay the blame for all its ills on the use of drift-nets by foreign vessels should be seen in perspective, government response on the issue has undeniably been far from adequate. ❐

situation was exacerbated by the granting — against the wishes of the country's administration — of fishing licences to two South African factory ships to operate off Namibia. This came at a time when, as a result of cautionary scientific advice, the administration had begun to stabilize quota sizes.

The allocation of these additional large quotas by South Africa to South African ships, made it impossible for the local management to resist industry pressure for higher quotas. The argument was, 'If we don't catch them, they will' — and the situation thus had all the makings of a commons tragedy.

The collapse started in 1969 after the 1968 peak of 1,5 million tons. By 1971 catches were down to 200 000 tons. At this point there was a scramble for remedial strategies, and the resulting stabilization of the catch was interpreted by the fishing companies as indicating that a recovery was at hand. The muddle of ad hoc directives culminated with the suspension in 1974 of scientific surveys, which were the only reliable way of estimating stock sizes. With quotas still set too high, the scene was set for the final nemesis of Namibian pilchard stocks. The 1976 season opened with a quota of 560 000 tons, which, as a result of the apparent scarcity of fish, was cut to 475 000 tons. In 1977 the quota was first cut to 250 000 tons, then to 200 000 tons, and finally, in 1978, to 125 000 tons. The landed catch in that year was 46 thousand tons.

There were a variety of factors which contributed to the depletion of Namibia's pilchard stocks. Besides the unsatisfactory collusion which developed between the authorities and business, to the exclusion of scientific advice, Namibia had the additional disadvantage of a colonial relationship with South Africa. The nature of this relationship as it affected the fishing industry, is well set out in the following quotation from an article by a fisheries scientist:

> It is possible that the basis for the decision to overrule the South West Africa Administration's objections, and to grant concessions, lay in a rapid decline of the South African pilchard fishing industry after a peak catch in 1963. This industry must have

needed a substitute fishing industry, and the seemingly stable expansion of catches in South West Africa would have been attractive. The way in which the concessions were granted has been cautiously criticized by the Commission of Enquiry and more openly in Parliament, where cooperation between fishing industry personnel and government was criticized as corrupt. It has been alleged that members of the government and public officials arranged for concessions to be awarded to companies in which they had financial interests.

It was of little comfort that anchovies filled the market gap left by pilchards. In South Africa, anchovy has traditionally been regarded as a fish of much lower value and has been unable to compete with pilchard as a canned product. It has instead been processed as fish meal, oil products and pet food. The change has entirely altered the economic and social structure of the affected coastal fishing communities.

Are there alternative management strategies?

The open access system is clearly not a useful strategy for ensuring sustainable yields. The limited entry system also has its limitations, not the least of which are the potential for political motives to override scientific advice. Moreover, there is the public perception, not entirely unjustified, of its being a euphemism for 'privileged access'. Clearly, demarcated property rights are not the solution for a mobile resource.

This leaves single entry — that is, nationalization — as the most serious contender for a long-term management strategy. While this may raise the spectre of inefficient parastatals and monopolies, the fact that it would be structured to maximize the yield from a common, renewable resource may preclude the usual productivity problems. It would also address the problem of state spending by taking away the need for a separate control function, and incorporating scientific personnel directly into the management structure.

A single company, with a single total allowable catch, could afford to optimize its operational activity in such a way as to maximize economic and ecological returns. Free from a web of vested interests, the company could obtain an objective assessment of projected maximum effort, and adhere to it without the inherent fear of competition. The integration of investment strategy with harvesting and processing capacity requirements would allow for a greater efficiency of operation, and greater long-term benefits.

Fishing is considered to be a high-risk industry. For this reason, fishing companies have a poor record of reinvesting profits into upgrading capital assets such as ships. In South Africa this is clearly shown by the fact that, in spite of high profitability, the average age of a fleet is older than the recommended retirement age of ships. Much of the profit goes instead into diversification, or to the holding companies. A single entry company could adopt a longer-term view, which would be in the interests of both the fishing industry and the beneficiaries.

Nationalization is the most serious contender for a long-term management strategy

FOCUS ON FALSE BAY

THE ARGUS

ETHENE ZINN

FALSE BAY is one of Cape Town's major attractions, and is intensively used by visitors and residents for recreational purposes. It is estimated that at the height of the season, more than 20 000 people a day visit its beaches. In the last few years, however, with the rapid urban and industrial expansion on the Cape Flats, it has come under severe threat as a result of the increased pollution load.

At present, pipelines discharge a minimum of 60 million litres of treated sewage, and 4 million litres of industrial effluent, onto the beaches or into the near-shore waters of False Bay on a daily basis. In addition, there are at least forty-one stormwater outfalls into False Bay. Six of these are major outfalls, and are reported to be highly contaminated with pathogenic viruses and bacteria. Some of them also have levels of suspended solids, nitrates, nitrites and lead that are considerably higher than generally accepted standards. Finally, a number of rivers entering False Bay are also badly polluted. The Eerste River, for example, carries run-off from agricultural land; sewage effluent from Stellenbosch, Kuils River and Macassar; and industrial effluents from the upper reaches of its catchment.

Although False Bay has a wide mouth, it is estimated that it has a water turnover time of between four and six days. Moreover, there is low exchange between the surf zone and the rest of the bay. There are already signs of a marked deterioration in the water quality of the bay, and with the prospect of even further development on its northern shores, this can be expected to worsen. Unless the situation is addressed soon, there are serious implications, not only for the ecology of the bay, but also for its value as a recreational facility.

Our coastal waters: surfer's paradise or cesspool?

Probably the most controversial 'use' of our oceans is for the disposal of large volumes of our society's wastes. This has the greatest potential to detract from, or impinge on, other 'uses' and thus the public interest. Although spectacular events such as oil spills attract more attention, about three-quarters of the pollution in the sea emanates from land-based sources. These sources include uncontrolled run-off from both urban and rural areas — which reaches the sea via rivers or storm-water drains — as well as pipeline discharges either into rivers, or into the sea itself. Illegal or accidental spills, such as the recent one from the Sappi mill in the Transvaal, also make a contribution.

Pipelines are the source over which we have the most control. South Africa currently discharges over 800 million litres of effluent directly into the sea every day, through sixty-one pipelines situated at various points around our coast. These pipelines operate under permits from the Department of Water Affairs, and discharge both industrial and municipal wastes. Disposal of these wastes is a problem and, in particular circumstances, a properly designed marine pipeline may be the best available option. However, on some parts of our coastline the pipeline effluents seem to be contributing significantly to the degradation of the receiving waters, which suggests that certain aspects have not been adequately considered.

The rationale behind disposal in the sea is that the huge volume of the oceans will dilute the waste to such a degree that it becomes harmless. There are, however, a number of factors which militate against such a simplistic approach:

- A number of South Africa's coastal towns (Cape Town, Mossel Bay, Port Elizabeth, etc) are situated on bays which have a limited exchange of water with the open sea, and which are therefore susceptible to a build-up of pollutant levels. Saldanha Bay, for instance, has an estimated water turnover time of twenty days. Nevertheless, about half of our sixty-one pipelines are situated in such bays. False Bay alone has ten pipelines discharging sewage, industrial effluent, or a mixture of the two. There is little doubt that these are contributing to the higher than normal nutrient levels recently reported for the near-shore waters of False Bay, which in turn could be linked to the apparent increase in the frequency of red tides there over the past few years. Seen together, these could be interpreted as the first indications of the onset of a gradual decline in its water quality.
- Research has shown that, rather than dispersing evenly throughout the sea, pollutants tend to be trapped or concentrated by physical features within the water body. For example, discharges close to the shoreline tend to be trapped in the nearshore waters. This finding has resulted in a shift from shoreline disposal to the construction of deep-water pipelines, which promote better dispersion. These are, however, very costly, and the majority still discharge their contents, as do rivers and stormwater drains,

ABOVE LEFT

Trek-netting at Strandfontein on the False Bay coast

ABOVE RIGHT

Children fishing on the famous pier at Kalk Bay harbour

Although spectacular events such as oil spills attract more attention, about three-quarters of the pollution in the sea emanates from land-based sources

inshore. The tendency for pollutants to be concentrated close to shore is reflected in the levels of the pesticide DDT, and the industrial chemical PCB, found in dolphins from the Natal coast. Bottlenose dolphins, which live and feed close inshore, have far higher concentrations of these chemicals in their blubber than common dolphins, which spend most of their time offshore.

• Finally, many of the more toxic pollutants — metals, pesticides and PCBs — are taken up into the tissues of marine organisms, where they can be concentrated to levels many thousands of times higher than those in the surrounding waters. For example, mussels have been shown to have levels of chlorinated hydrocarbons up to 690 000 times higher than the surrounding water. Apart from the possibility of harming the organisms themselves, these accumulated substances represent a potential health hazard to anyone consuming them. Filter feeders, such as mussels and oysters, can accumulate not only chemical pollutants, but also pathogens such as the bacteria and viruses present in sewage. Unacceptably high levels of pathogens have, for example, been found in mussels in the vicinity of the Paapenkuils canal in Port Elizabeth.

What are the alternatives?

In the first instance, industries should be required to reduce as far as technologically possible the amount of waste produced. Experience has shown that, although there has often been initial resistance to these ideas, in most cases the industries concerned have been able to come up with innovative concepts and have ultimately increased their own profitability. For example, organic wastes can potentially be used to fertilize other forms of growth. In the United States, new legislation forced pulp mills to look at their waste in a more creative way. The result was that a valuable yeast used as a cattle feed additive was grown in waste ponds, clarifying the water in the process.

In addition, alternative methods of disposal should be investigated, not only to reduce disposal to the sea, but also within the context of an improved water management policy. (See Chapter 9.) In a country like South Africa which has limited water resources, it is ludicrous that, for example, 60 per cent of the water used by the city of Cape Town, is used for the transport and disposal of sewage.

Where marine outfalls are appropriate, they should be situated in deep water, on open coastlines. Moreover, their siting should take the other uses of the body of water concerned into consideration. Water quality criteria have been recommended for the various uses to which coastal waters are put, those for bathing and shellfish harvesting being the most stringent. Pipelines should only be considered where it is certain that they will not detract from the required water quality. Finally, allowance should be made for public participation in all planning decisions, including the siting of marine outfalls.

Pipelines, however, are not the only source of marine pollution.

Of greater concern, perhaps, is the pollution coming from illegal, accidental, and 'uncontrolled' sources. Not only is there little real idea of exactly what volumes are entering the sea from these sources, but it is going to take a great deal of effort and ingenuity to establish effective controls.

Another source of marine pollution in South African waters is shipping. Some 150 million tons of oil a year are transported past our coastline from the Middle East. The average size of the tankers is 250 000 tons, so that the potential for large and dramatic spills is significant. Long-term environmental damage, though, is more likely to be a result of smaller, but more frequent, deliberate discharges. In the last few years the incidence of these has apparently declined, possibly due to the use of patrol aircraft and the institution of higher fines. South Africa is, actually, relatively well equipped to deal with oil spills — no doubt a consequence of the publicity they attract.

Coastal waters are being polluted either for short-term profit or as a result of the casual attitude of the authorities

Obviously what is necessary now is to raise the profiles of other sources of pollution, and to make polluting more costly for the culprits.

Where to now ?

Clearly the public interest in the sea and seashore is being compromised on a number of fronts — if not in terms of apartheid legislation, then as a result of politically-based policies which favour private and/or business interests. Parts of the seashore are being alienated for private developments, with the sale of land no doubt generating large sums of money for developers or local authorities. The public are also excluded from mining areas. Fishing industry companies have made huge profits by over-exploiting common resources, but have failed even to provide job security for their employees, or any substantial benefits for the rest of the community. At the same time, there is a host of regulations covering the collection of fish or shellfish by individuals.

However, the public interest extends beyond the question of access. There is little advantage in having access to over-exploited resources, or a degraded environment. While the situation is not beyond salvation, there is no room for complacency either. Coastal waters are being polluted either in the interests of short-term profits, or as a result of the casual attitude of the authorities. This is exacerbated by government ineptitude of a different kind, namely the lack of adequate housing, which has led to the establishment of huge squatter areas. Estuaries are being degraded by the profit-motivated activities of, for example, sugar farmers. There is a general lack of appropriate control over development in the coastal zone.

Clearly a comprehensive set of policies, addressing all of the above issues, is needed if we are going to retain the benefits provided by the oceans for all of our people, and to fulfil our responsibilities to the international community. Concerted action by ordinary South Africans to lay claim to their heritage is just as crucial.

•NAN RICE•

When Nan Rice sees a family of dolphins or whales peacefully swimming in Cape waters she can be forgiven if she takes some credit for the fact that they are there — alive, well and protected from harassment. For twenty years, she has devoted a great chunk of her life to the drive to ensure safety for these magnificent sea creatures.

The process has been a learning experience not only for Rice, but also for the South African public and government officials; and this education has been achieved through the work of the Dolphin Action and Protection Group (DAPG), with its threefold aim: education about, protection and conservation of whales, dolphins and porpoises.

All this happened by accident. Although growing up in the then undeveloped coastal area of Sea Point gave Rice a love of nature, and although animals were always one of her passions, she had always believed that her helping skills would be directed towards people rather than wildlife.

It was while living in Hout Bay as a wife and mother who was involved in assorted welfare activities, that Rice had the experience that changed her life. Summoned by her frantic children, who had witnessed about 200 dolphins being hauled out of the early morning surf in a net, she arrived at the beach to be greeted by a scene of carnage. 'They were just being grabbed and dumped. I saw about four die in front of me — it was the most appalling scene I've ever seen.'

Her horror was balanced by a characteristically pragmatic response. 'I realized what I had seen was news.' Aided by statements and photographs from a group of visitors from Johannesburg, Rice went to the press with her story. It was published, but the response was far from satisfactory. She was told she didn't know what she was talking about.

'I was beaten down with patronizing male statements. I was terrified, but I held my ground.' Today she is not terrified, she knows exactly how to handle the media and, she claims, many of the things she said in the early days are now being said by scientists.

That was 1969 and dolphins in South African waters had no protection at all at the time. 'They were being exploited, people were catching them for dreadful little pools.' In 1970, Rice prevailed upon the then Administrator of the Cape, Nico Malan, to promulgate legislation protecting dolphins in the Cape from killing, capture and harassment.

In 1973 the Sea Fisheries Act was passed, extending the protection to the rest of South Africa's waters. Yet the slaughter continued as permits were granted to 'people who shouldn't have had them'.

But awareness was growing and the Dolphin Action and Protection Group, with Rice as its secretary, was founded in May 1977, its motto and policy being, 'Dolphins should be free'.

In 1979, with the first Save the Whales campaign in South Africa, the DAPG's activities were extended to include the protection of whales. In 1980 the group presented a dossier to the government on the international trade in killer whales and whales were drawn into the protective net of legislation.

Inevitably, Rice's involvement in and expertise concerning the issues grew. Information gathered by the group on pirate whaling was incorporated into reports submitted to the International Whaling Commission and the United States Senate Enquiry.

Papers written by members of the Group's Dolphin Whale Watch project have been presented at international conferences. From its small start in one woman's reaction to the carnage on Hout Bay Beach, the group has grown

and its activities have spread. It has linked up with organizations such as Greenpeace in a Save Antarctica campaign and it is also involved in campaigning against the plastic pollution which is damaging the world's marine environment.

More recently, the DAPG has taken up the struggle against another international abuse, that of gill-netting, with its wholesale plunder of the seas for food and the attendant massacre of non-edible marine life randomly caught up in the massive nets. The group is also working towards the formation of a great white shark protection and research unit.

There is no challenge related to the protection of sea life that Rice will not take up. A no-nonsense, hyperefficient woman, she has turned what might have been no more than an emotional issue into a series of highly-organized campaigns which have paid off in tangible results.

'Politicians make many of their decisions on a politico-economic basis and do not take into consideration conservation efforts. It's a good thing to have a non-governmental organization to keep them on the straight and narrow.'

Though DAPG does not have more than 200 members at any one time because it prides itself on all its members being active, it has many friends — hundreds of people are, for instance, involved in the issue of plastic pollution and that, too, requires coordination.

Much of DAPG's work is done through education, with a barrage of literature being distributed countrywide to schools, tourists resorts, game parks and the public at large. Lobbying is a priority.

Each issue Rice and her group tackle has broad-ranging ramifications and implications.

The gill-netting outcry, for instance, had an entirely unexpected spin-off when the focus on the Taiwanese tuna fishermen who were robbing South Africa's waters revealed the appalling conditions under which their South African crews were working. The result is that Rice has become a consultant to the Food and Allied Workers Union who, in turn, have become allies in the battle against gill-netting.

She attributes much of her success to her skill as a campaigner. She believes campaigning is an art. 'There are ways and means. You must have the aims and objectives sorted out before you campaign and if you stick to them it's very easy. You decide what you are going to do, produce the literature and then you must know exactly when is the right time to say things, to lobby.'

It's a hectic life but one senses she wouldn't swap it for another. Apart from anything else it would be very difficult to extricate herself. 'I am just caught up in the vortex and I whirl around. You belong to the public and have very little personal and private life but when you have made the commitment it carries a lot of responsibility. When I look back over twenty-one years, I can see that I've achieved an enormous amount with the group. We have credibility all over the world — the word has definitely spread.' ❐

14

THE CRUCIAL LINK

Conservation and development

◆ MARGARET JACOBSOHN ◆

Conservation in context

Elephants returned to water at the remote pro-Namib spring of Purros in early 1987 after an absence of nearly ten years. This sign that the desert-adapted wildlife of the Kaokoveld was at last recovering after the worst drought in living memory, combined with the devastating poaching of the 1970s, was welcomed by conservationists. Not so the residents of Purros.

'Shoot the elephants. Or tell Nature Conservation to take them away. This is a place for people: we don't want wild animals here.' Kata Tjambiru, owner of one of the carefully tended mealie, pumkin and watermelon gardens at Purros, spoke for most of her community when she expressed these hostile views the day after a nocturnal visit by the elephants. This time they had not entered the thorn-branch-enclosed gardens, but they would be back; and as everyone knew, elephants in a field of crops could destroy months' worth of food in minutes.

A Herero elder who had lost most of his cattle in the 1979-1982 drought kicked the fresh elephant droppings near the gardens and recalled with bitterness the last months of the drought: 'When my cattle were starving Nature Conservation chased them out of the Skeleton Coast Park — the last place where there was still food. They said that area was for wild animals and they would shoot cattle that came in. I had to put my cattle inside a kraal and watch them die, knowing that just downriver, inside this park, there was fodder.'

The man closed his eyes, remembering his once vast herd, collapsing in twos and threes to die the slow death of starvation. At first men had attempted to lift the gaunt animals back onto their feet — it took four men and two stout wooden poles — but soon there were too many for this. He opened his eyes: 'So why doesn't Nature Conservation keep its elephants away from our food?'

His wife spoke up: 'Today I am married to a poor man. I must sweat in these fields to grow a few mealies because our cattle are now too few to give our children milk. And families no longer can live together. After the rains women must stay here to tend the gardens while our men take the stock far into the hills. Our days have become dry and heavy. Now we have also elephants to worry about.'

Her husband's younger brother nodded: 'That's straight talking. The wild animals are Nature Conservation's cattle. They must put them behind a fence and keep them away from our place.'

Until that morning, after spending only a few weeks living with the people of Purros, I had never seriously questioned western conservation practices and ethics in Namibia or anywhere else. From my comfortable, well-fed, urban vantage point, I firmly believed that all nature conservation was good and admirable and that conservationists were doing wonderful work. Not only were they preserving wild areas for future generations and myself to enjoy, but they were ensuring biodiversity in the world's various ecosystems, rescuing endangered species from the brink of extinction and fighting a war against poaching.

Living at Purros gave me uncomfortable new insights into tourism, that benefactor of conservation

My months with a community of semi-nomadic Himba and Herero herders in the Kaokoveld drastically changed my views: I learnt what it feels like to lie inside a frail dung-plastered dwelling without a door and listen to a lion roar nearby: on such nights I would recall the silver scar tissue marking the forearms of a Purros man who had fought a lion that attempted to enter his hut. He survived only because his neighbour owned an old .303 rifle. I learnt that while a rhino or elephant is an exhilarating sight from a four-by-four truck, it can be terrifying for an unarmed woman walking through thick riverine vegetation to fetch water at a spring. I worried alongside Himba and Herero mothers when the children who had taken the goats out were late returning in the evening. My initial response, that no mother should allow young children to go out alone in the Kaokoveld, was soon tempered by the realization that the children were essential labour — there was simply no-one else available to herd the smaller animals. I listened attentively as an old man explained to his youngest grandson, aged about five, what he should do if he met lion, leopard, hyena, rhino or elephant when out goat-herding. I shared the anger and distress of families who lost kidgoats to jackals, sometimes two or three a week. And when one of the rare Skeleton Coast lions killed a cow, I empathized with the man who had owned her, even though I could not fully share his shock and grief. Many times during my stay in the Kaokoveld, I found myself thinking that Nature Conservation's priorities seemed both arrogant and unjust.

Living at Purros also gave me uncomfortable new insights into tourism, that benefactor of conservation. In short, I learnt that from a Himba or Herero person's perspective, nature conservation and tourism inevitably meant varying degrees of hardship, personal danger, inconvenience and above all, social and economic disempowerment. In Namibia in the 1970s an area could be declared a

'Shoot the elephants. Or take them away. This is a place for people. We don't want wild animals here'

— Kata Tjambira

national park with no real community consultation or any compensation for the herders who had used it for centuries in times of drought; in the 1980s a tour operator could be given a concession to lead tours or build a tourist camp in an area without the local people even knowing, let alone having a say in the matter. To be fair, all blame cannot be laid at the door of the Directorate of Nature Conservation and Tourism. The so-called second tier ethnic authorities set up by central government were responsible for decisions about matters such as tourism concessions (from which they received benefits). But such ethnic authorities did not always consult at grassroots level, nor did they necessarily represent all the people in their 'ethnic' group.

If one looks at it from the point of view of many ordinary rural Africans, it comes as no surprise to learn that wildlife conservation is failing across much of the continent. All too often the demands of the western conservation ideology conflict with the legitimate

GARTH OWEN-SMITH

needs of usually impoverished rural communities. Small wonder that such communities, who may live within a key conservation area or alongside national parks or game reserves, view wildlife and tourism with at best, indifference, and at worst, outright hostility.

The challenge facing this generation of conservationists, therefore, is not only to reconcile wildlife conservation priorities with the needs and aspirations of Africa's people, but also to link their economic and social development to conservation. If we fail to do this, big game will become extinct in the wilds, no matter how many crack anti-poaching squads with firepower and helicopters we have in the field, no matter how many ivory or rhino horn trade bans we support.

The concept that wildlife conservation should involve and benefit local rural communities is not new, and in fact, in the last few **years it has become part of mainstream conservation philosophy.**

BELOW
Elephants return to Purros in the Namib Desert

But in most areas we are only starting to grapple with the actual implementation of such strategies. We urgently need to find and implement practical ways of channelling conservation-related benefits back to rural communities and, equally important, to involve them in the management of their natural resources. People whose relationship with their natural environment has been disrupted are unlikely to behave responsibly towards that environment. In other words, what is required is the re-empowerment of people so that they can control their own resources and therefore their own lives.

What follows is an outline of a small integrated rural development-conservation project that aimed to link conservation and tourism to a community's socio-economic development. It was set up in late 1987 in Purros in north-western Namibia by the people of Purros, freelance nature conservationist Garth Owen-Smith, and myself.

The project has used conservation and tourism to broaden an impoverished community's economic base, thereby positively influencing the people's attitude to wildlife. It has also attempted to counter some of the negative effects of tourism. It has built upon and expanded the effects of the auxiliary game guard system which was set up by conservationists and Damara, Herero and Himba community leaders in 1983.

The Purros Project

The background to the project

Purros is a strong, sweet spring that rises from the bed of the lower Hoarusib River about 50 kilometres from the Skeleton Coast. Purros and the surrounding area includes some of the most starkly beautiful mountain and desert scenery in southern Africa. With its permanent water supply, it is a key area for game, including the now famous desert-adapted elephant and black rhino.

Most of the area's game was wiped out by poachers in the 1970s. Poaching was brought under control in the early 1980s when non-government conservation bodies, the local people and government conservationists began cooperating.

'When my cattle were starving Nature Conservation chased them out'

— Herero elder

By 1987 all wildlife populations were increasing in the Kaokoveld. About sixty semi-nomadic Himba and Herero smallstock and cattle herders used Purros as their base at that time. Like others in the west, they were among those herders hardest hit by the drought. Across the Kaokoveld, formerly wealthy cattle breeders lost an estimated 80 to 90 per cent of their largestock. Up to 130 000 head of cattle and tens of thousands of sheeps and goats died of starvation around the territory's few permanent springs before the drought broke early in 1982. The people survived thanks to emergency feeding schemes set up by the then South West Africa/Namibia Government, the South African Defence Force, Red Cross International and local service organizations such as Rotary.

When the drought ended most Himba families decided to return

MARGARET JACOBSOHN

ABOVE
A woman from Purros milks her cow

to their semi-nomadic herding lifestyle and set about rebuilding their herds. But without sufficient cattle to rely on milk as a staple, their economic independence was lost and for the first time, the Himba people had need of an economy beyond their own pastoral-subsistence one. Formerly spurned, wage labour, the cash economy, mass-produced goods and schooling all gained ground as never before.

Perhaps the greatest disruption of all was the war between the South African Defence Force and the South West African People's Organization (SWAPO). Many Himba lineages including some of the families now living at Purros, fled west and south to avoid being embroiled in the conflict. Young men whose family herds had been depleted by the drought found well-paid work in the army, thus becoming the first generation of Himba wage-earners.

When I started my research at Purros in 1986, several years after the drought had ended, the community were still impoverished and were struggling to build up their meagre cattle herds to their former numbers. Families no longer had enough cattle to meet their subsistence needs throughout the year. While there was enough milk after the rains when grazing was plentiful, people went hungry in the late dry season, which could last from November to March.

Vegetable gardens in alluvial silts below the spring in the Hoarusib bed, initiated as a self-help project by a well-meaning ser-

In the past nature conservation and tourism have inevitably meant varying degrees of hardship and social and economic disempowerment

vice organization at the height of the drought crisis, provided an alternative food supply. But for semi-nomadic herders, the furrow-irrigated gardens were socially and ecologically disruptive. Horticulture in this terrain requires daily labour and therefore, a sedentary population, whereas successful pastoralism demands a mobile, semi-nomadic settlement pattern. This meant that families had to split up. Throughout the year the older but still active women and a few small children remained at Purros to tend the gardens. They needed milk so a few cows and some goats were left behind with them. This stock tended to graze the riverine vegetation, going against the general herding strategy which was to conserve fodder around permanent springs until as late as possible in the dry season.

Following the pattern of wild grazing animals, Himba herders traditionally move away from permanent waters as soon as it has rained. Taking advantage of all surface water and local rain-related pasture, the herders keep away from their permanent springs until they have no further option. The riverine and other vegetation thus conserved then supports the stock in those last hot, dry months until the rains come, and the semi-nomadic cycle starts again. A permanent presence of stock at Purros was therefore most undesirable, but at the same time, the gardeners who had consumed milk as their staple food all their lives, could not be expected to do without it for months at a time.

Tourism

During the 1980s the Kaokoveld experienced increased tourism. With the availability of four-by-four vehicles, more and more

RIGHT
A tourist takes pictures of the Purros settlement

guided and unguided safaris were made into the once remote terrain. Purros, with its permanent spring, spectacular scenery and 'traditional' Himba homesteads was, and still is, a popular destination.

While many tourists were courteous visitors, on the whole tourism's impact at Purros was largely negative: the breathtaking rudeness of some tourists who, for example, photographed Himba people without asking permission or even greeting them, caused amazement and then anger. As a result, all tourists began to be viewed with suspicion, and then treated as fair game for handouts — tobacco, scraps of food, sugar, sweets, medicines and so on. But as these tourist 'payments' inevitably went to those prepared to pose for photographs, some families started camping near the road so as to be first in line. Their stock stayed nearby, further undermining the essential semi-nomadic grazing practices. Social disruption also resulted as families who received nothing from tourists became resentful of those who did: the sharing ethic that tied families into relationships of mutual reciprocity rapidly started unravelling under the onslaught of tourism.

At the same time, increased contact with westerners created an appetite for western foods and goods, but no-one at Purros was engaged in wage labour to meet these new needs. People thus resorted to asking tourists for goods, a practice they saw, in accordance with their own social customs, as conferring status on the tourist. Tourists, on the other hand, interpreted the request along western lines as begging, and therefore as demeaning.

Thus in Purros the situation was fraught with potential for a mutually degrading relationship.

By late 1987 Purros was in a state of crisis, and each new party of tourists divided the community still further. Accusations of witchcraft were being made; people were ill from spells they believed jealous neighbours had cast; nomadism was breaking down; the community was hostile towards the returning elephants and towards a lioness which was in the area. People were also disturbed by rumours that a tourist camp was to be built at Purros and that the area would be inundated with strangers.

The inception of the project

At this point we sat down with the people of Purros to seek solutions. Our aim was to ask how we could reconcile conservation priorities and tourism with the aspirations and needs of the community in practical ways.

At that first meeting, attended by most of the people of Purros, Garth Owen-Smith and myself, we discovered that few, if any, people saw the connection between the region's wildlife and the many tourists now visiting the area. And although wildlife in this magnificent desert setting was undoubtedly the region's major marketable resource, none of the local residents perceived it as such.

The existing benefits from tourism — random handouts — gave tourists and tour operators demeaning power over the local people and generated social tensions. These handouts in no way ade-

GARTH OWEN-SMITH

The contact between tourists and the community was fraught with potential for a mutually degrading relationship

quately compensated the people for damage, danger and inconvenience that lion, rhino and elephant might cause. The people made the point very forcibly that they knew whites did not have to put up with lion and other dangerous animals on their farms, so why should they? Moreover, they pointed out, there were plenty of young SADF-armed Himba and Herero soldiers around to deal with any so-called problem animals.

We realized as we listened at that meeting that if we wanted active support for conservation of game from the Purros people, meaningful and tangible benefits had to be generated by conservation. For all its negative aspects, the expanding tourist industry seemed the only short-term way of passing much-needed material benefits from wildlife to the struggling Purros community. But a more businesslike arrangement between tourists and residents that would promote mutual respect and generate direct local benefits to the whole community was necessary. At the end of a two-day meeting, a pilot project that involved a tourist levy, a craft market and a 'pool' of workers for conservation-related jobs, was born.

The tourist levy

Each tourist spending a night at Purros as part of the eight-day Koakoveld tour organized monthly by the Endangered Wildlife Trust (EWT) and guided by Garth Owen-Smith would pay the Purros community a levy of R25 for the use of their communal land and for the privilege of seeing the wild animals that shared it with them. This levy would be paid directly to the Purros community as a whole and would remain separate from any area concession fees paid to regional or national authorities.

A major tour operator in the area, Schoeman's Skeleton Coast Fly-In Safaris, joined the scheme in 1989, agreeing to pay the community R5 for each tourist who visited Purros. (Schoeman's tourists do not stay overnight at Purros as do the EWT parties.)

An immediate problem facing us at Purros was that no institution existed to receive and redistribute cash on behalf of the community. The people had to create a suitable body and the first Purros Conservation Committee was chosen. After several hours of discussion and argument the meeting made the democratic decision that all adult men and women deserved a share of the tourist levy: it was agreed that six lineage heads would serve on the committee. This overruled a suggestion by some of the younger men that the levy be shared out among all adult men. The women at the meeting vehemently opposed this idea, preferring to entrust the levy to the six older, responsible family heads. That way, as one matriarch put it, women would have a say in how the money was used and the whole family would benefit, not just the men.

At that stage the community voted to give each of the six lineages an equal share regardless of family size. In retrospect, we realized this was a wise decision because it allowed the divisions at Purros to heal, precluding acrimonious discussion about how much each family should get. Two years later the people enlarged the committee to eight and decided that while the levy should still be shared between families, the size of that share should depend on the number of adult men and women per family.

The people also decided that every family should receive their share of the levy, even if they did not come into contact with tourists. All agreed that this would ease the tensions that had developed between families and encourage the reestablishment of normal semi-nomadic movement in the area. This has in fact happened in spite of warnings from opponents of the project that it would attract people to Purros to seek a share of the levy. This so-called magnet effect has not occurred for the simple reason that neither we (Owen-Smith and myself) nor the tour operators have any say in deciding who deserves a share. This has been left in the hands of the people themselves and of course, there is little chance of them allowing outsiders to muscle in.

BELOW

A Purros resident making goods for the craft market

MARGARET JACOBSOHN

The committee also serves as a mouthpiece for the Purros people, liaising with tour operators and others. The community thus have some say in how tourism is conducted in their area: they know more or less when tour groups will be visiting and they have told tour operators how they would like groups to behave. For example, tourists are asked not to walk across the sacred fire area in homesteads.

The craft market

The second part of the Purros project involved the setting up of a craft market to sell Himba and Herero goods. Initially, people found it hard to believe that tourists would pay money for handmade items but today most women and some of the men routinely make articles such as baskets, wooden pails, neckrests, jewellery and belts — whenever they need cash. At first, as Purros is 101 kilometres from the nearest trading store at Sesfontein, I undertook to take orders and buy required goods on my regular supply runs to

Although wildlife was undoubtedly the region's major marketable resource, none of the local residents perceived it as such

Khorixas, Swakopmund or Windhoek. Most of this function has now fallen away as the community, assured of a regular cash income, has arranged for periodic visits from a Sesfontein trader.

Work in conservation

The third aspect of the project involved the setting up of a casual labour pool for conservation-related work. The community decide who has the skills required for a particular job. For example, a young Herero woman who has six years of schooling has been hired to lead a count of the palm trees whose fronds are used for basket-weaving. She and her team spend a few days each year on this task so that tree numbers can be monitored against over-exploitation. At her request she is paid in cash and in mealie-meal. Men are hired as trackers during regular rhino-monitoring operations; they request a daily cash fee plus food while in the field. At present these ad hoc payments are made from funds raised by non-government conservation bodies.

By late 1990, when the project had been running for three years, the tourist levy, the craft market and ad hoc employment had generated more than R25 000 for the people of Purros. These benefits are directly related to wildlife and its conservation as the community is now well aware that tourists come into the area primarily to see the game. Understandably, a formerly hostile community now has a very positive attitude to wildlife and to tourism.

THE AUXILIARY GAME GUARD SYSTEM

The innovative and successful network of game guards in communal areas in Namibia was the brainchild of Garth Owen-Smith and Damara, Herero and Himba community leaders in 1983. Although it played a pivotal role in stopping the slaughter of elephant, black rhino and other game in the Kaokoveld, it should not be confused with an anti-poaching unit which attempts to catch poachers: the aim of the auxiliary network has been to stamp out poaching by involving and empowering local communities in the conservation of their wildlife.

Unlike members of anti-poaching units, who are appointed by government and non-government conservation organizations, the auxiliaries are always appointed by their own community leaders and remain responsible to them. This promotes community accountability for natural resources while at the same time forcing government conservationists to enter into a partnership with communities. As a Kaokoveld headman noted with satisfaction: 'Nature Conservation officers now come to us to talk, not just to lift the lids off our cooking pots seeking game meat.'

In view of Africa's colonial history, it is essential for conservationists to work towards a changed relationship with communities in their areas of jurisdiction. The auxiliary network attempts to give back some real power to communities.

GARTH OWEN-SMITH

LEFT
A black rhino in the Kaokoveld

An old lineage head remarked on receiving his levy: 'It is as if we are farming wild animals. But instead of getting meat and skins for them, we get the money that the tourists pay to see them. That is why we must look after our wildlife.'

Another lineage head reported his own son after the young man had poached a zebra. With quiet dignity, he told Garth Owen-Smith: 'Let him be charged in court. I will pay my son's fine by selling a beast from my herd.' When I asked him why he had turned in his son, he said: 'My family agreed that we would not kill our wildlife and Himba people do not go back on their word.'

'My family agreed that we would not kill our wildlife and Himba people do not go back on their word'

— Himba lineage head

Some benefits of the project

An important aspect of the craft market, tourism levy and employment is that they provide ways for people to earn an income without having to abandon their subsistence economy to squat at the edge of an army base or town in search of scarce or non-existent wage labour. The project thus gives these rural people access to the cash economy. It has acted as a bridge, teaching people who were outside or on the edge of the cash economy the cash value of the goods they sell and buy.

It has also taught people to value and take renewed pride in their own skills: a skilfully woven milk basket can earn its weaver R60 or more, and this buys a large, good quality blanket or a big cooking pot.

Another effect of the project has been a renewed interest in wildlife, and heightened respect for the older generation who lived closer to nature than people do today. The older people's vast stores of knowledge and experience about wildlife and its habits have been revitalized and made accessible to the young. As a good cash wage goes to trackers hired to monitor particular species, young men are now keen to learn about wildlife from their elders.

In 1987 the traditional skills, such as basket-weaving and wood carving, and knowledge about wildlife, were regarded as irrelevant by young people. Today people of all ages see these skills and knowledge stores as giving them access to a more secure future.

Tourism is now on a more dignified footing at Purros. Local people feel they have some control over how tourism takes place and they therefore meet tourists as hosts and equals. Most tourists paying the levy have responded to the idea with enthusiasm: there is a world of difference between the paternalistic experience of a tourist who hands out a packet of biscuits or some sweets in return for a posed picture of ochred Himba women in calfskins, and the tourist who meets Purros people knowing that all present, including him or herself, are partners in conservation.

There can be little doubt that from a social and economic point of view, the Purros community has benefited from the project. Conservation has also progressed; in 1990 the Directorate of Nature Conservation selected Purros for a restocking operation involving twenty-nine gemsbok and ten giraffe from Etosha. This was the first time game had ever been translocated from a national park into

People of all ages now see traditional knowledge and skills as giving them access to a more secure future

a communal area. Christopher Eyre, senior nature conservationist in charge of the area, asked the lineage heads of Purros to take care of the gemsbok and giraffe. 'The game will be all right around Purros; we work well with the people here,' he said.

Although tailored to the unique socio-economic situation at Purros and involving a relatively small community in a remote and wild terrain, the project has direct relevance in areas where local people are more urbanized and where population pressures are much greater. A few basic principles apply, whether one is working (in conservation or tourism) with semi-nomadic pastoralists in Namibia, among over-crowded communities dependent on both subsistence agriculture and wage labour in KaNgwane or KwaZulu, or among commercial sheep farmers in the Cedarberg. These are:

- dialogue
- shared decision-making in matters of common interest
- joint planning from the grassroots level
- cooperative action

There will be no benefits without involvement and empowerment. Nor can integrated community development-conservation schemes be designed in nature conservation head offices: the local community must be involved from the start. Both conservationist and community will obviously have to make compromises along the way, but it is important that there is real dialogue; that all decision-making is shared; that local people take part in all planning and that action involves them as well as conservation authorities.

Into the future

Where does the Purros Project go from here? At a 1990 post-independence workshop to restructure the Directorate of Nature Conservation, Namibian conservationists selected the principle of involving local communities and returning benefits to them as one of their top five priorities. It is hoped that the principles embodied in the Purros Project will become part of the new national Namibian conservation and tourism policy. Ideally, tourism-generated levies should be paid in all communal areas where there is wildlife.

Meanwhile, at Purros people no longer see conserving wildlife as preserving an irrelevant part of their past. Now that conservation is firmly tied to the community's economic and social development, wildlife has become something that they value. If people across Africa could be similarly influenced, Africa's wildlife would also have a future.

15

ANIMALS VERSUS PEOPLE

The Tembe Elephant Park

◆ THE ASSOCIATION FOR RURAL ADVANCEMENT (AFRA) ◆

The Tembe Elephant Park is situated in northern KwaZulu, along the Mozambique border, in the district of Ingwavuma. It was the first game reserve to be established by the KwaZulu government in 1983, and provides some insights into the problems of conservation in Africa.

There were a number of African families living in the area who had lived there for generations and who had adapted their lifestyle to the dry environment. The people of Ingwavuma are extremely poor; they practise a subsistence agriculture and harvest local resources, but are dependent on migrant labour earnings for their survival.

A sketch map of Ingwavuma, showing the major reserves

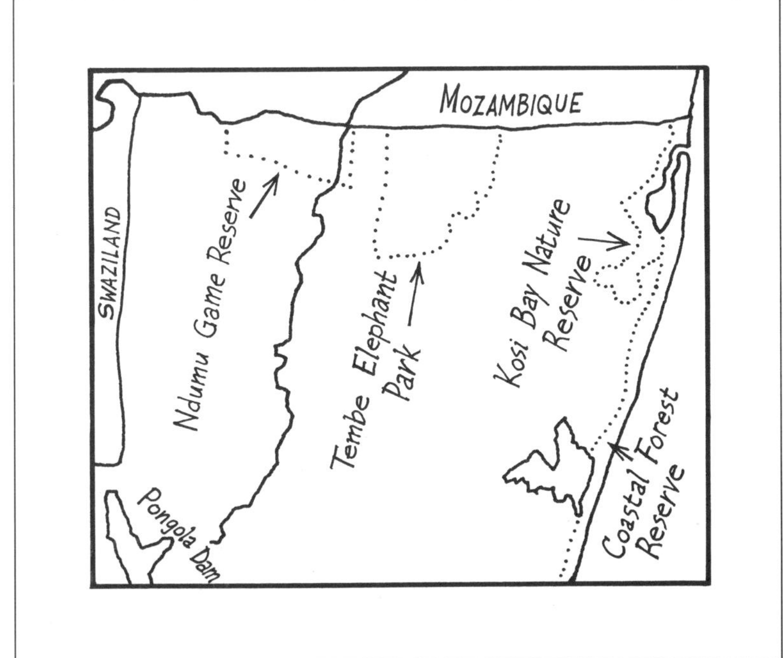

The creation of the Tembe Elephant Park was followed by the removal of thirty-two homesteads (KwaZulu's figure) by the newly established Bureau of Natural Resources (hereafter referred to as the Bureau). Some of the people who were moved maintain that there was no consultation about the move. They were relocated outside the boundaries of the Park where there was no water supply. By 1989 some were still complaining of inadequate compensation for the loss of homesteads, harvests and cattle.

Forced removals and social dislocation have often accompanied the establishment of nature reserves in Africa

Similar forced removals and social dislocation have accompanied the establishment of game and nature reserves elsewhere in Africa since colonial times. In this case the legacy of such practices is a deep-seated hostility amongst many people in Ingwavuma towards the Bureau's conservation plans, such as the creation of the Kosi Bay Nature Reserve in 1988.

Conservation will not successful in the long term unless it can survive political change, and this can be guaranteed only if conservation projects have the support of the local people. Gaining this support is not easy, for there will always be conflict between long-term versus short-term goals. But means must be found if South Africa's natural heritage *and* human heritage are to be equally protected.

A lack of consultation

The history of conservation in South Africa is riddled with projects that were set up without consulting the indigenous people. Yet it is these people who have been most affected by conservation. Although the policies of the KwaZulu Bureau of Natural Resources are an advance on 'colonial' conservation policies, the Bureau's

practices are leading to similar results.

According to local informants, no formal meeting was held to discuss the fact that the local people would have to be moved once the Tembe Elephant Park was proclaimed: the notification that the chief wanted them to move from the area (ie on instructions from tribal authority), came through a third party. In addition, although the people affected were opposed to the move, they say that there was no meeting or opportunity for expressing their opposition.

Local people do say that at some stage a white Bureau official called a meeting to say that the area was going to be made into a game reserve. However, there was apparently no induna or tribal authority official present, and nothing was said about forced removals. When later the fence was being constructed, the people were told by Bureau officials that they must move.

The irony is that in the previous year, senior KwaZulu officials had publicly opposed the threatened removal of Chief Zikhali's people from Sodwana Bay when the latter reserve was excised from KwaZulu and handed over to the Natal Parks Board. Chief Minister Mangosuthu Buthelezi is reported to have criticized certain South African government departments for forced removals in the past, and to have said that this merely proved 'white greed' and the fact that blacks were just 'pawns' in South Africa (*Daily News*, 3 September 1981; *Natal Mercury*, 5 June 1981).

Recently, the Chief Minister has claimed that KwaZulu will not establish game and nature reserves at the expense of local people. But the experience of the people of the Kosi Bay Nature Reserve — proclaimed in 1988 — throws some doubt on this claim.

BELOW
Women gather reeds for housing near Kosi Bay in Natal. Local inhabitants have been in conflict with the Department of Parks and Conservation since the gathering of reeds was criminalized recently

The removal from the Tembe Elephant Park

In late 1988, a woman who is head of a family of six children, said: 'We were moved during the ploughing season and left cultivated fields as well as marula and *umnwebe* trees [which are harvested for wild fruit and food supplements] behind.' (*Weekly Mail*, 28 October 1988). Others complained of losing cattle in the move. They were all removed to sites outside the Park fence.

Compensation

The Bureau has claimed that all the people evicted from the Park were fully compensated.

A couple who were resettled told AFRA in 1987 that they were given poles and rafters to rebuild their homes, but that these were insufficient. They were also given between R60 and R80 by the Magistrate of Ingwavuma, but the purpose for which this money was intended was never explained to them.

The Chief Induna of the Tembe Tribal Authority, Mr Solomon Tembe, is reported to have said in late 1988 that some of the relocated people still had not been compensated (*Natal Witness*, 10 December 1988).

By September 1989, at least three people who had been removed from the Park were complaining to AFRA that they had

The old 'colonial' idea that conservation cannot be practised unless indigenous people are removed no longer holds water

not yet been compensated for the loss of homesteads and fields. This was despite being referred from the Tembe Elephant Park to Ingwavuma, and from Ingwavuma to the tribal authority office at Manguzi/Ngwanase.

It is KwaZulu policy to channel 25 per cent of tourist-related revenue to the tribal authorities for the benefit of the local people. By late 1988 (six years after the eviction), the people removed from the Tembe Elephant Park claimed that they had seen no benefits because the Park was still closed to the public. The Bureau confirmed this but said that a tourist camp was being planned, and once this was complete, the local people would benefit materially.

In fact, the revenue will be given to the local tribal authority and there is no guarantee that the people affected by the creation of the Tembe Elephant Park will benefit.

The relocation site

The major complaint about the new sites was the lack of water. A local woman said in 1988: 'We were promised a water supply but never got this' (*Weekly Mail*, 28 October 1988). She said that they had to walk long distances daily to collect water from an abandoned quarry. The Chief Induna, Mr Tembe, confirmed the lack of water in December 1988 and added that there was a rumour that the Bureau was at one stage thinking about pumping water into the Park to top up the pans for the elephants.

The Bureau admitted that it was studying the feasibility of providing artificial water in the Park, but said it was willing to allow people outside the fence access to the water inside the Park. A Bureau representative said that the water issue was the responsibility of KwaZulu Department of Water Affairs. He blamed the tribal representative for failing to take up the problem with the local authorities.

The Bureau has also claimed that it permits the local people to harvest natural resources from conservation areas. However, in 1988, women adjacent to the Tembe Elephant Park were complaining that although they were allowed into the Park to collect plants and thatch, their men were forbidden entry, apparently to prevent poaching. The Bureau tacitly admitted this, but said that this had recently been rectified in negotiation with the local tribal chief.

Human hardships

Thus, in the interests of protecting wild animals, a number of people who had lived in the area of the Tembe Elephant Park for generations were forcibly removed, without proper consultation.

They were deposited outside the boundary fence where there was no water supply, and the promised access to the Park for harvesting natural resources was initially curtailed. They waited five years before the first sign of financial gain from tourism was given, and by 1989 certain people claimed that they had still not been compensated for the loss of homesteads, fields and cattle.

This explains why the people of Ingwavuma are suspicious of the Bureau's conservation policies, and why the creation of the Kosi Bay Nature Reserve in 1988 has evoked so much hostility. The local people cling to their rural lifestyle and their land. One of the resettled people said: 'We are not happy that our place was given to the elephants. I feel sore that the elephants are there at the place of our ancestors. There were always elephants when we were staying at the old place, and if they came near the home, we would chase them away' (AFRA file, 1987).

Conservation can be guaranteed in the long term only if it has the support of the local people

Future removals

There is a narrow corridor of land — the Mbangweni corridor — between the Tembe Elephant Park and the Ndumu Game Reserve (established by the Province of Natal in 1924). In 1988 there was a report that the Bureau would link the two reserves. Various estimates of the number of people who are likely to be affected if these plans proceed range between 1 164 and 3 000 people.

Conservation and development

The old 'colonial' idea that conservation cannot be practised unless indigenous people are removed no longer holds water. A large body of traditional conservation practices exists amongst the people of Ingwavuma and this knowledge can form the basis for the development of sound management policies. But the local people must be consulted and directly involved in conservation planning and practices.

Furthermore, conservation must be linked to rural development so that the material and social interests of the indigenous people are protected. If they are offered viable alternatives to practices which are ecologically destructive, for instance, then conservation need not cause hardship for human beings.

Conservation will not gain the approval of the local people if it continues to lead to forced removals and material deprivation.

The danger is that without local support, and in an era of political change, the long-term future of game and nature reserves in KwaZulu could be in jeopardy.

•IAN PLAYER•

All the major religions of the world, says Ian Player, were founded by people who went into the wilderness — 'the source, the fount of inspiration which has literally changed the world.'

This is the reason why Player has, for more than thirty years, spoken out in South Africa for the wilderness. As founder of the Wilderness Leadership School, Player, part mystic, part psychologist, part practical game ranger, has led literally thousands of people on trails through the Umfolozi, introducing them to the physical wonders of the wild and to the spiritual experience of wilderness.

In all that time, he has never lost the sense of wonderment which forms the basis of his philosophy and the foundation of his belief in the restorative value of the wilderness experience.

He has found ancient example in the Celts (from whom, through his mother, he is descended); religious precedent in Moses, Christ and Muhommed; psychological validation in the work of Jung; and a mentor in that literary and literal venturer into the interior, Sir Laurens van der Post. But essentially, when he talks, when he writes and when he leads parties on well-loved trails, the experience he imparts is his own.

It is an experience which can, does, change lives. It changed his. When as a young man he returned to this country after fighting in Italy during the Second World War, it was to find a South Africa locked into the post-war depression. A stint underground, digging for gold, was only one of the ways in which he earned his living.

In 1952, he joined the Natal Parks, Game and Fish Preservation Board and took his first steps into the wilderness. He did it — as he expects those who follow him to do it today — on foot, on horseback and in a canoe. He was the first person to paddle the route of what is now the Duzi canoe marathon. He was captivated by the whole experience.

'I know now,' he says, 'the psychological process that was taking place.' It is a process explained by Jung, one which has been acknowledged through the centuries. 'One of the Elizabethan poets, Sir Thomas Browne, said, "We carry within us the wonders we seek without us: there is all Africa and her prodigies in us." We evolved in Africa, so the wilderness experience in Africa is literally coming home. The unconscious is mirrored by the external, the two things come together and you can never be the same.'

This is Player's philosophical base. He quarrels with those who believe he is a fanatic but concedes that he has 'fought desperately hard for the wilderness'.

That is not surprising. After all, it was he who in 1955, after being introduced to the wilderness concept by a colleague, Jim Feely, persuaded the Natal Parks Board to set aside two small areas — the Umfolozi and Lake St Lucia — 'where one can walk, ride and canoe away from the sights and sounds of human artefacts.' It was a move, he believes, which changed the whole environmental movement in South Africa.

'It has established another form of consciousness and understanding of what the environment is all about. As the Americans have put it in their wilderness legislation: wilderness is 'the environment of solitude'.

Even today, when the concept of wilderness has gained currency, it is a constant problem to maintain selected areas as true wilderness; to prevent seekers after twentieth century comforts from putting down roads and driving vehicles along them. It is, says Player, 'an enormous struggle, but there is an energy and an impulse towards an understanding of the environment and our place in it.'

Sadly, he observes that the growth in environmental consciousness is not a growth in wilderness consciousness. He believes, though, that wilderness areas 'are the new temples, the places where people go to be healed.'

Though his beliefs are rooted somewhere way

back in the origins of the world, Ian Player is not one who would wish to turn back the clock on development. 'Science is enormously important in the modern world,' he says. 'It's the new magic, but it isn't the only thing, it has also destroyed a lot. You have to make room for the other side, which people glibly term "emotional".'

The compulsion to give people the experience of wilderness, which led to the establishment of the Wilderness Leadership School, springs from an overwhelming gratitude which Player has always felt. So it is that the comparatively tiny area of Umfolozi has played host to more than 100 000 people, 'experiencing Africa through the souls of their feet'.

Though Player has personally accompanied about 2 000 of them and has given them the benefit of his experience, he does not see himself as a teacher.

It is the wild that is the teacher: the sights, the smells, the feel of the bush; the sounds of the night beyond the small camp fire; the special peace. 'If you are a good trails officer you let the wilderness do the work. I don't want to talk and I don't want to hear them talk.'

Looking back on a longish life during which he has travelled the world on lecture tours, written innumerable words and turned thousands of city people into lovers of the wild, Ian Player believes 'my wilderness work has been my chief achievement — it's enabled other people to discover more about themselves.' And that to Player is possibly the most valuable gift of all. 'In the second half of life you begin to realize the only journey that's important is the journey inward.'

But how does he see the future of that exterior world in which human and animal are interdependent and human and nature are inseparable? That's a little hard to say: he is not committing himself, though he does believe that 'one person can change things' and that that in itself is hopeful.

He does warn, though, that 'either you work with nature or she will knock you out and she'll do it across the board'. And since he is quite fond of both humans and animals, he believes it would 'be a great tragedy if we become an extinct species'.

Then again, he thinks that's quite unlikely. He has a firm belief that 'there is this ancient goddess within us all and she's coming back. There is a return to the recognition of the feminine, of the earth as the ultimate mother. There's nothing soppy or sentimental about that, it's real.'

Getting down to basics, Player acknowledges that it's going to take a little more than a belief in ancient truths to keep the planet safe. Pragmatically, he says, 'the world is in too bad a state for the green movement to fail. Today we are citizens of one world and a lot will depend on the next generation and the one to follow as to how people will live.

'Everybody has a responsibility ... to ourselves, to our children and grandchildren and, above all, to the earth, and it is one earth. I think the message is there and the world has at last begun to realize that if the earth dies, we die.'

A bleak prospect and not one he dwells on for too long. After all, he has invested a lifetime in making people care about the fate of the earth and the flora and fauna living in it. Player is one of the few who can look at his small corner of this beleaguered planet and say, 'I've done my best and will continue to do so.' ❐

16

BIODIVERSITY: THE BASIS OF LIFE

South Africa's endangered species

◆ JOHN LEDGER ◆

Ecosystems comprise a multitude of species which in turn are made up of assemblages of genes. These three components collectively make up biodiversity — the biological diversity that provides the basis for life on earth, including that of humans. Threatened species

RIGHT
A roan antelope — one of the endangered species in South Africa — is fitted with a radio collar in the Pilanesberg National Park

may often indicate biodiversity at risk, and many species are known to be in danger of extinction in South Africa. There are compelling ethical and practical reasons for protecting the biodiversity of South Africa, but historical and socio-political considerations could prevent this from becoming a national priority in the short term. This chapter argues that the conservation of South Africa's biological diversity should be prominent on the agenda for a new socio-political dispensation.

Human concern for the welfare of the planet is a relatively new concept in world history. The first international organization devoted to this cause was l'Office International pour la Protection de la Nature, founded in Brussels in 1934 through private initiative. This body evolved into the International Union for the Protection of Nature (IUPN), and in 1956 the name was changed to the International Union for the Conservation of Nature and Natural Resources (IUCN), now referred to as the World Conservation Union.

The IUCN initiated the idea of compiling lists of threatened wildlife as a means of drawing the attention of decision makers and land managers to the plight of species faced with extinction. These lists became known as Red Data Books (RDBs), and the first internationally applicable one was published in 1966. The IUCN also encouraged individual states to draw up lists of threatened species within their own political boundaries.

JOHN LEDGER

The value of biodiversity

Biological diversity implies the variety of the world's organisms, including their genetic diversity and the assemblages they form. It is an umbrella term for the natural biological wealth that supports human life and well-being. Genes are the building blocks of species, and species are the components of ecosystems.

Species are central to the concept of biodiversity. Individual species — plants, animals and microorganisms — provide the maize and meat we eat, the antibiotics doctors use to save lives, and numerous other natural products. Threatened species are symptoms of ecosystems under stress; they can therefore be used as indicators of ecosystem health. Threatened species are those which are treated as endangered, vulnerable or rare (as defined by the IUCN) in international or regional RDBs, or other authoritative sources.

Biological resources provide the basis for life on earth. The social, ethical, cultural, and economic values of these resources have been recognized by humans in literature, religion and art from the beginning of recorded history. More recently, the governments of the world represented at the United Nations made an important ethical commitment to nature. The World Charter for Nature was adopted by the General Assembly on 28 October 1982, and it expresses unqualified support by governments of the principles of conserving biodiversity. However, this charter has almost been forgotten by governments and conservationists.

The World Charter for Nature recognizes that humankind is part of nature, that every form of life is unique and warrants respect regardless of its worth to human beings, and that lasting benefits from nature depend upon the maintenance of essential ecological processes and life-support systems and upon the diversity of life forms. It calls for strategies for conserving nature, scientific research, monitoring of species and ecosystems, and international cooperation in conservation action.

AN ETHICAL BASIS FOR CONSERVING BIODIVERSITY

The International Union for the Conservation of Nature and Natural Resources has produced an ethical basis for conserving biodiversity, based on the World Charter for Nature and the IUCN World Conservation Strategy of 1980:

- The world is an independent whole made up of natural and human communities. The well-being and health of any one part depends upon the well-being and health of the other parts.
- Humanity is part of nature and humans are subject to the same immutable ecological laws as all other species on the planet. All life depends on the uninterrupted functioning of natural systems that ensure the supply of energy and nutrients, so ecological responsibility among all people is necessary for the survival, security, equity, and dignity of the world's communities. Human culture must be built upon a profound respect for nature, a sense of being at one with nature and a recognition that human affairs must proceed in harmony and balance with nature.
- The ecological limits within which we must work are not limits to human endeavour; instead, they give direction and guidance as to how human affairs can sustain environmental stability and diversity.
- All species have an inherent right to exist. The ecological processes that support the integrity of the biosphere and its diverse species, landscapes, and habitats must be maintained. Similarly, the full range of human cultural adaptations to local environments should be enabled to prosper. Sustainability is the basic principle of all social and economic development. Personal and social values should be chosen to accentuate the richness of flora, fauna, and human experience. This moral foundation will enable the many utilitarian values of nature — food, health, science, technology, industry, and recreation — to be equitably distributed and sustained for future generations.
- The well-being of future generations is a social responsibility of the present generation. Therefore, the present generation should limit its consumption of non-renewable resources to the level that is necessary to meet the basic needs of society, and ensure that renewable resources are nurtured for their sustainable productivity.
- All persons must be empowered to exercise responsibility for their own lives and for the life of the earth. They must therefore have full access to educational opportunities, political enfranchisement, and sustainable livelihoods.
- Diversity in ethical and cultural outlooks towards nature and human life is to be encouraged by promoting relationships that respect and enhance the diversity of life, irrespective of the political, economic, or religious ideology dominant in a society.

There is a growing tendency to place economic values on the maintenance of biodiversity. This is not easily done, but McNeely et al (1990) state that

> '... in order to compete for the attention of government and commercial decision-makers in today's world, policies regarding biological diversity first need to demonstrate in economic terms the contribution biological resources make to the country's social and economic development. Even partial valuation in monetary terms of the benefits of conserving biological resources can provide at least a lower limit to the full range of benefits and

demonstrate that conservation can yield a profit in terms that are meaningful to national accounts.'

Three main approaches have been used for determining the value of biological resources:

- Assessing the value of nature's products, such as firewood, fodder and game meat, that are consumed directly, without passing through a market ('consumptive use value')
- Assessing the value of products that are commercially harvested, such as timber, fish, game meat sold in a market, ivory and medicinal plants ('productive use value')
- Assessing indirect values of ecosystem functions, such as watershed protection, photosynthesis, regulation of climate, and production of soil ('non-consumptive use value'), together with the intangible values of keeping options open for the future ('option value') and simply knowing that certain species exist ('existence value')

The loss of a species from the web of diversity can have far-reaching, perhaps catastrophic, results

The greatest value of biodiversity lies in the genetic material contained in the species that make up the ecosystems. It is impossible to attach a price-tag to this, but without the genetic building blocks, the future viability of life is threatened. The loss of a species from the web of diversity can have far-reaching, perhaps catastrophic, results. When a species becomes extinct, it takes with it the hard-won lessons of survival encoded in its genes over millions of years. Only about 1,7 million of the estimated 30 million life forms on the earth have been catalogued, and the way things are going, hundreds of thousands of them may be extinct by the year 2000.

The food base of humankind depends mainly on three wild grasses — wheat, maize and rice. These species have been refined and developed to provide huge yields, but they are now very vulnerable to disease. Crop scientists are constantly turning to the wild relatives of domesticated plants for a source of new genetic attributes. The loss of these wild plants would remove the vital material that humankind needs to keep its food base in good shape.

One third of all active ingredients in western medicines are based on plant compounds. Examples are aspirin, derived from the willow tree, digitalis, quinine, morphine, the drugs used to treat leukemia, and countless others. Yet only 2 per cent of the earth's 300 000 species of flowering plants have been properly investigated for pharmacological compounds. At the same time, these species are being destroyed faster than they can be studied.

Global threats to biodiversity

McNeely et al (1990) have compiled a comprehensive overview of the global threats to biodiversity, summarized here.

Humans entered the industrial age with a population of one billion, and with biological diversity (the total of genes, species and ecosystems on earth) at a probable all-time high. Biological resources were freely available for exploitation to support development.

Species are being destroyed faster than they can be studied

In the late twentieth century, with the world population approaching six billion and still growing, it is now recognized that biological resources have limits, and that humans are exceeding those limits, thereby reducing biological diversity. Each year more people are added to the human population, species are becoming extinct at the fastest rate known in geological history, and the planet's climate appears to be changing more rapidly than ever.

Human activities are progressively eroding the earth's capacity to support life, at the same time that growing numbers of people and increasing levels of consumption are making ever greater demands on the planet's resources. The combined destructive impacts of an impoverished majority struggling to stay alive, and an affluent minority consuming resources to excess, are inexorably and rapidly destroying the buffer that has always existed, at least on a global scale, between human resource consumption and the planet's productive capacity.

Deterioration of the planet's life-support systems is likely to continue until human aspirations adapt to the realities of the earth's resource capacities and processes, so that activities become sustainable over the long term. The problems of conserving biological diversity therefore cannot be separated from the larger issues of social and economic development.

Maintaining maximum biological diversity assumes far greater urgency as rates of environmental change increase. Diversity in genes, species, and ecosystems provides the raw materials with which different human communities will adapt to change, and the loss of each additional species diminishes the options for both nature and people to adapt to changing conditions.

Biodiversity in South Africa

Many aspects of the concepts and conservation of biotic diversity in southern Africa have recently been reviewed in an excellent book by Brian Huntley, Clem Sunter and Roy Siegfried (*South African Environments into the Twenty-First Century*, 1989).

Because of its position at the tip of the continent, and the presence here of the winter rainfall region which hosts the Cape floristic kingdom, South Africa is one of the world's richest areas of biodiversity. McNeely et al (1990) have listed the top ten African countries with the highest numbers of species for selected organisms. South Africa is sixth for mammals, with 279 species, ninth for birds with 725 species, sixth for amphibians with 93 species, first for reptiles with 281 species, and first for flowering plants, with 21 000 species. The latter figure is twice as high as for the countries ranking in the next three places, Zaïre, Madagascar and Tanzania, with 10 000 each. On a global scale, South Africa is ranked sixth in the world for the number of its flowering plants.

In southern Africa (including Namibia and Botswana) there are an estimated 23 200 plant species, of which 18 560 (80 per cent) are endemic to the region. This gives the area the highest species richness in the world (calculated as species per area ratio). Of the southern African plants, 2 373 are regarded as threatened, 1 621 of which are in the Cape floristic kingdom. This gives this region the highest concentration of threatened plants of any temperate region (McNeely et al, 1990).

More than 560 national parks and nature reserves, totalling over 7,2 million hectares, have been set aside to protect South Africa's diverse fauna and flora. Nearly 97 per cent of the birds, 93 per cent of the mammals, 92 per cent of the amphibians and 92 per cent of the reptiles are included in this protected area network, which makes it one of the most effective systems of its kind in the world (Siegfried, 1989). The fact that much of this protected area network has its roots in the colonial and apartheid eras should not be allowed to detract from its value. This rich diversity of landscapes and wildlife is the basis of South Africa's tourism industry, which was valued at approximately R3 150 million in 1987 (Huntley et al, 1989), and which has considerable growth potential. The value of this industry will be vitally important to future citizens of South Africa.

Overlying the rich biological tapestry of South Africa is the most intensive industrial, agricultural and urban development on the African continent, in an area dramatically impacted by human activity (Macdonald, 1989). The fact that South Africa produces 60 per cent of the continent's electricity is but one measure of the degree of this development. Other chapters in this book provide details of the utilization of natural resources in South Africa. The Witwatersrand, Durban and Cape Town are some of the fastest-growing urban areas in the world.

South Africa is the only country on the continent to have produced a series of national Red Data Books (SARDBs). Five volumes have been published by the Council for Scientific and Industrial Research (CSIR) as South African National Scientific Programmes Reports. The SARDBs deal with plants (1980; 1985), birds (1984), terrestrial mammals (1986), fishes (1987), reptiles and amphibians (1988) and butterflies (1989). Future revisions of the existing SARDBs, and new titles in the series, will be produced jointly by the Foundation for Research Development (FRD), an independent offshoot of the CSIR, and the Endangered Wildlife Trust, a non-government member of the IUCN dedicated to the protection of biological diversity.

Ferrar (1989) provides a useful review of the value and the shortcomings of the SARDBs. Their purposes are to identify rare and declining species, to establish the nature and extent of such rarity or decline and to suggest priorities for conservation, monitoring and research. The concept of protecting habitat to protect rare or declining species has actually been the reason for the establishment of major southern African protected areas in the past (Ferrar, 1989).

The SARDBs have played an important role in public awareness, in research and monitoring, and in planning and administration. Their principal weaknesses are that they generally contain sparse and subjective data, they provide an inadequate scale of res-

When a species becomes extinct it takes with it the hard-won lessons of survival encoded in its genes over millions of years

The combined destructive impacts of an impoverished majority struggling to stay alive and an affluent minority consuming resources to excess are destroying the planet

olution, and they tend to place excessive emphasis on rarity, which Ferrar (1989) perceptively identifies as 'an ecologically respectable condition'.

One of the useful aspects of RDBs is that they can help to identify 'indicator species' that point to the ecosystems or habitats on which they depend being under stress. Conversely, if a particular organism is thriving, we can usually assume that the pyramid of life which supports it is also in good health. This is a useful tactic for protecting genetic diversity, since time is not on our side. Some information from the SARDBs is given below, to illustrate the value they have in identifying threats to biodiversity.

Flora

Two plant RDBs have been published (Hall et al, 1980; Hall and Veldhuis, 1985). Hall (1989) gives an overview of rare plant surveys and atlases. The value of computer data banks is particularly important, and two are being developed. They are the Cape Inventory of Critical Environmental Components (CICEC) and the National Atlas of Critical Environmental Components (NACEC). The threats to the plants of South Africa are particularly serious, and RDBs are not able to fulfil the same role that they do with groups of animals. Ferrar (1989) has described the situation as follows: 'Plant taxonomists, understaffed, underfunded, and locked into a painfully time-consuming set of procedures, stand little chance of ever reaching their goal of a complete Regional Flora at the present rate of progress.'

Many authors have pointed out that the world is losing species faster than they can be described, and their potential value to ecosystems and humankind investigated. South Africa is no exception. To illustrate this point, consider the effects of rapid urbanization in Cape Town on the most threatened temperate flora in the world.

The Cape Town metropolitan area has grown dramatically, and the 1980 population of 1,9 million is expected to double by 2010. Urbanization has been particularly intensive on the Cape Flats, with massive housing schemes being developed for homeless people. Only 1,6 per cent of the Cape Flats flora is afforded protection inside formal reserves. The areas favoured for housing development, close to the sea and on flat ground, coincide with the distribution of Strandveld and coastal fynbos. Some endangered species such as *Jordaaniella dubia* and *Euphorbia marlothiana* occur only here. The area of the Cape Flats which links Muizenberg to Milnerton has seventy-five species of plants that are listed as endangered. This area of about 100 square kilometers has the highest concentration of endangered plant species in the world (Schneier, 1990).

Hilton-Taylor and Le Roux (1989) state that of the total area of the fynbos biome, 14 per cent is conserved, most of it in the mountainous areas. The lowlands of the south-western Cape have the highest priority for conservation attention. The priorities within these vegetation types have been established, and precise sites ranked according to a number of criteria. A detailed conservation guide plan has been drawn up, yet the areas continue to be degraded by urban and agricultural development. The same authors rate the conservation status of the succulent Karoo (1,2 per cent conserved) and the Nama-Karoo (0,5 per cent conserved) as poor in comparison with that of the fynbos. These two biomes are inadequately studied, but the need for conserving their biodiversity is very important. Hilton-Taylor and Le Roux (1989) make the point that improved utilization and management of the Karoo shrublands is more important in the long term than the establishment of protected areas.

JOHN LEDGER

ABOVE

The only viable South African population of wild dogs is found in the Kruger National Park. This animal is the most threatened carnivore in Africa

Terrestrial mammals

Of the three endangered species listed, the wild dog is today found only in the Kruger National Park, with a small and possibly non-viable population in Zululand. This species is incompatible with livestock ranching, and has been completely eliminated from farming areas. Its conservation status is unlikely to improve significantly, and the role of the Kruger National Park in the survival of this animal is therefore crucial. The roan antelope was largely eliminated from grassland areas that were suitable for agriculture. Today its status is improving significantly as a result of the game farming industry and its economic value (around R30 000 per animal in 1990), and it will most likely be off the endangered list in the next edition of the RDB. The riverine rabbit is a victim of the

The fact that much of South Africa's protected area network has its roots in the colonial and apartheid eras should not be allowed to detract from its value

general degradation of the Karoo and the agricultural utilization of its habitat. Some protected areas are being established on private land, and the species has every chance of increasing, now that the threats to its survival have been identified (Smithers, 1986).

Birds

Five species are listed as endangered, and two of them have proved to be particularly good indicator species for two specific biomes.

The wattled crane has been reduced to about 120 breeding pairs, most of them in Natal, and the balance in the eastern Transvaal. The birds require a large, permanent 'sponge' or vlei on which to breed. They are vulnerable to disturbance by humans and dogs. Wetlands in South Africa have been severely degraded by draining, ploughing and burning, and the wattled crane is a sensitive indicator of wetland quality. In this regard it is also a 'flagship species' that can be used to heighten public awareness about the need for wetland conservation.

The blue swallow is restricted to the montane grasslands of the eastern escarpment, from Natal through Swaziland to the eastern Transvaal. A survey in 1985 recorded only sixty-three breeding pairs throughout its range. The immediate reason for the demise of the blue swallow is the loss of habitat caused by the conversion of montane grassland to agriculture, and in particular, the planting of exotic trees for South Africa's burgeoning commercial timber industry. It is regarded as South Africa's most endangered bird (Brooke, 1984).

Reptiles and amphibians

Three endangered amphibians are recorded from the southern Cape, two of them on the Cape Flats (Cape platanna and micro frog) and one on Table Mountain (Hewitt's ghost frog). All are threatened by extensive urban development and associated disturbance of the environment (Branch, 1988). Although the South African herpetofauna has been intensively studied and is considered to be rather well known, a new genus and species of frog from the south-western Cape Province, *Poyntonia paludicola,* was described as recently as 1989 (Boycott, 1990).

Fishes

There is a major concentration of threatened endemic freshwater species in the south and south-western Cape, particularly in the Olifants River system. The threats to the freshwater biome here consist mainly of the extraction of water for agriculture and industry, the damming of watercourses, and the spread of exotic vegetation (Skelton, 1987).

Overview

Despite various shortcomings, South Africa's biodiversity, and the threats to it, have been well described by numerous authorities, and in a series of RDBs. No other country in Africa has succeeded in achieving a comparable state of knowledge. There is thus a wealth of information and expertise in this country that could ensure the

conservation of biodiversity more successfully than would be the case in most other countries in the world. However, historical and socio-political factors could impede this endeavour.

The colonial powers had no ecological consideration when demarcating states in southern Africa. Later, the architects of apartheid carved South Africa into various independent states and self-governing 'homelands'. The resulting patchwork of autonomous conservation authorities has cut across biomes, river catchments and other natural environmental divisions. Some rural areas have suffered severe reductions in biodiversity, especially those places where large numbers of subsistence farmers with their livestock and dogs have been crowded onto land unable to sustain them. Other areas are today in pristine condition because property ownership laws have denied access to impoverished people.

Modern South African society represents a microcosm of the global environmental scenario. Its human population comprises an affluent minority and an impoverished majority, the former with a low birth rate, the latter growing rapidly because of a high birth rate. Due to their close juxtaposition in South African society, the aspirations of the impoverished are very often to emulate the affluent. The fact that such aspirations are both largely unattainable, and unsustainable, will come as a disappointment to many hopeful people in South Africa.

The allocation and management of resources in the recent past of South Africa has been dictated by authoritarian dispensations that started with the colonial era and continued through the apartheid era. With the end of the latter in sight, and with a new order about to emerge in South Africa, the demand for resources will increase dramatically. One of the most pressing issues is access to land for dispossessed and homeless people (Kahn, 1990a). Land set aside for the conservation of biodiversity may be perceived as

ABOVE
The Egyptian vulture is one of three birds to have become extinct in South Africa in recent times. The others are the African skimmer and the yellowbilled oxpecker — the latter has now returned

The fact that the aspirations of the impoverished majority are both largely unattainable, and unsustainable, will come as a disappointment to many hopeful people

'unproductive', and therefore available for human occupation, or it may be considered to have previously been inhabited by people whose descendants now have a right to reoccupation.

Regrettably, biodiversity as a meaningful concept is masked for many people (probably the majority) by two perceptions: first, that nature conservation is a Eurocentric and colonial concept that has no place in the future, and second, that endangered species are large, hairy, or irrelevant creatures (like rhinos) which whites regard as more valuable than black people (Kahn, 1990b). This last notion has been promoted by certain writers seeking any rod with which to beat the old order in South Africa.

This chasm is lamentable, for there is an urgent need to create some kind of consensus among all South Africans concerning the environment that must sustain this and future generations, irrespective of what political dispensation may prevail. It is vitally important that the biological diversity of South Africa be regarded by all its citizens as a priceless asset that deserves the highest respect and protection the nation can provide.

The road ahead

The current status of South African society is at variance with one of the most important clauses of the IUCN's Ethical Basis for Conserving Biodiversity, notably:

> All persons must be empowered to exercise responsibility for their own lives and for life on earth. They must therefore have full access to educational opportunities, political enfranchisement, and sustaining livelihoods.

Until South African society fulfils these criteria, it could be difficult to persuade its citizens that the conservation of biodiversity should be a national ethical concern. Nevertheless, every attempt should be made to do so during the transitional period towards a democratic South Africa.

The preoccupation of some whites with wildlife preservation at the expense of, for example, dispossessed rural communities may be historically demonstrable — but this should not blind us to the fact that South Africa now has one of the best systems of protected areas anywhere in the world. This is a national treasure from which all future South Africans will benefit. Local expertise is helping to save the black rhino for future generations, having already done the same for the white rhino. These are significant achievements on a global scale, and all South African citizens should be able to share in the pride of such accomplishments.

A sense of perspective is required, one in which all South Africans see themselves as part of the global family. This could be promoted by the various political organizations endorsing the World Charter for Nature as part of their manifesto. A global strategy for conserving biodiversity is currently being developed by a coalition of the World Resources Institute, IUCN, and UNEP (the United Nations Environmental Programme) in close collaboration with the World Wildlife Foundation, Conservation International, the World

Bank, the Asian Development Bank, and other key government and non-government institutions in both tropical and temperate nations (McNeely et al, 1990). This aims to

- establish a common perspective, foster international cooperation, and agree to priorities for action at the international level
- examine the major obstacles to progress and analyse the needs for national and international policy reform
- specify how conservation of biological resources can be integrated with development more effectively and identify the linkages with other related issues facing humanity
- promote the further development of regional, national, and thematic action plans for the conservation of biological diversity, and promote their implementation

In addressing the debate about land, it will be important for South Africa to join the world community in trying to ensure that adequate areas of the various biomes are set aside to protect their biodiversity. The advice of Siegfried (1989) in this regard is relevant: he states that

> Given that a common principle goal of the nature conservation agencies in southern Africa is the long-term conservation of a maximum of biological diversity in nature reserves, it is not yet too late for devising and implementing an innovative strategic plan for realizing this goal. In this plan, the present nature-reserve system should not be regarded as sacrosanct. If and where necessary, parts of, or whole reserves should be deproclaimed. The land could either be sold or exchanged in order to obtain new areas needed for conserving plant and animal associations that are under-represented in the current system.

A sense of perspective is required, one in which all South Africans see themselves as part of the global family

Some conclusions

The 1990s may be the last decade during which constructive and creative decisions, activities and investments — rather than emergency rescue efforts — can be made to ensure that many of the world's species and ecosystems are maintained, examined for their material and ecological value, and promoted for sustainable use to support new and innovative approaches to development. The combination of maintaining the maximum possible biological diversity, the maximum possible cultural diversity, and the greatest possible scientific endeavour would seem the most sensible approach towards the dynamic challenges facing humanity (McNeely et al, 1990).

The world is at a crossroads in the history of human civilization, while South Africa is at a crossroads in its political history. Our actions in the next few years will determine whether we take a road towards a chaotic future characterized by overexploitation and abuse of our biological resources, or take the opposite road — towards maintaining great biological diversity and using biological resources sustainably. On a global scale, the future wellbeing of human civilization hangs in the balance. At the local level, South Africans should see themselves as part of that human civilization, notwithstanding the injustices of the past that have made this difficult to believe at times.

•AILEEN TAYLOR•

An anti-vivisectionist who doesn't align herself with animal lovers? Why should Aileen Taylor, industrial sociology masters student, be fighting for the right of animals not to be tortured in the interests of medical science or vanity?

She is, it seems, being true to herself and true to the philosophy with which she was brought up. 'I am a respecter of all life forms with a deep reverance for all that is living, rooted in a sense of justice, non-exploitation, and a holistic approach to the environment,' she says.

Many of those she refers to contemptuously as 'bunny huggers' or 'traditional animal lovers' eat meat and 'if it came to an issue of my baby or my dog would still condone experiments on animals', she avers. Taylor does not eat meat.

Her unsentimental attitude to animals stems from her growing-up years in Pretoria. 'We had quite a few animals — bantams, budgies, dogs — and there was always an attitude that you treated animals well; maltreatment was something very strongly objected to. My father has quite a deep sense of compassion for animals and was quite a nature child in his youth.'

Taylor's respect for the environment grew as she attended girl guide camps, enjoying the outdoors, thriving in the camping environment and becoming aware of the need to protect the world around her. During her years at Natal University in Durban, she learnt other things that made an impact upon her ever-searching mind.

She dipped into social activism, feminism and, while she was tutoring, was introduced to anthroposophy, a subject which was to become enormously important to her. She defines it as 'the study of the higher intellectual states of the human being'.

It is a discipline which leads, she says, to questions about the 'whole role of human existence within the universe and your place in the universe and why it is that the relationship between human and non-human animals is such an alienated, mad, unhealthy one lacking a spiritual aspect and a sensitivity towards understanding the right to life.'

With her growing abhorrence of exploitation and any form of abuse, it was not unexpected that Aileen Taylor would, sooner or later, become involved in the problem of exploitation of animals. She sees vivisection as a symptom of the continuous power struggle in which the human race appears to be engaged.

Though her academic life has nothing to do with her activist life, there too she is exploring questions of power, looking at forms of resistance and control in the workplace and using the garment workers of the thirties and forties as her case study.

The issue of vivisection did not enter her consciousness until shortly before she left Durban. Her introduction came through a friend who convinced her that the practice was not, as she had thought, 'for the general good of society'. An exhibition of horrific pictures put the visual seal on her friend's verbal picture and Taylor was hooked into a campaign which has radically altered the shape of her life.

At the launch of the Durban branch of Earthlife, Taylor picked up some literature and began informing herself. She was shaken again when she saw a documentary on animal experimentation and began, she says, to realize for the first time 'that experiments on animals were something that humans don't need'. From that point it was only a matter of time before she became unshakeably opposed to all animal experimentation, unwilling to concede that any use of animals can be justified.

She insists, though, that her stance is not an 'anti-human ethic'. Vivisection simply does not benefit humans, she maintains. 'People think they need to have their medications tested on animals, but it has been proved that experiments on animals often yield different data and that much of the evidence hasn't been supported.'

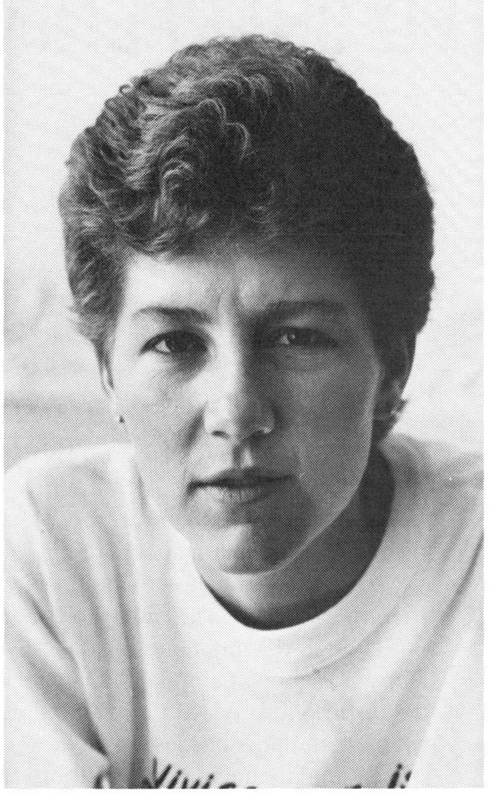

The initial focus of her conversion was the World Day for Laboratory Animals in April 1990. In the run-up to that period, she read extensively and produced a fact sheet. With some amazement, she found interest in the campaign growing both in the media and among ordinary supporters.

The result was the formation by a group of activists of an organization called South Africans for the Abolition of Vivisection (SAAV). SAAV immediately held its first public campaign, a picket on the public road near the Roodeplaat Research Laboratories outside Pretoria.

The first surprise on that April day was the fact that about 200 people arrived to demonstrate. They included lawyers, scientists, doctors and a church minister. The second surprise was the decision of the demonstrators to march on the laboratories. A somewhat bemused Taylor found herself coordinating the march.

The demonstration attracted enormous publicity and the anti-vivisection cause was placed firmly in the consciousness of the public. SAAV's next major move was a march on the animal unit of the University of the Witwatersrand, where Prof Graham Mitchell, head of the unit, was declared Vivisector of the Year 'for his dedicated defence of vivisection in the belief that human ailments can be cured by studying non-human animals.'

Taylor is very aware that the anti-vivisection lobby has traditionally been seen as hysterical and espousing an anti-human ethic. A new movement has recently been born, though, opposing vivisection purely on medical and scientific grounds, and it is with this movement that which she aligns herself.

'It's not a moral issue — there's clear evidence that the methodology is inaccurate and scientifically fraudulent.

The medical or scientific response tends to take the argument further. The more you debate morality or try to define what is necessary and unnecessary, the further you get away from the evidence which will refute animal experimentation,' she contends.

Taylor sees the issue as having ramifications for the entire South African population. 'Health care is a major issue and that's why it's a political thing, and it's perturbing that so much money is being spent on animal research laboratories when it should be spent on health facilities, especially in South Africa.

'It is safe for people to have vivisection abolished, I really believe that.' ❐

17

THE TEETH NEED SHARPENING

Law and environmental protection

◆ JON WHITE ◆

The law has not played a significant part in protecting the South African environment to date. Most environmentally aware people would say that the reason for this is inadequate environmental protection legislation — but few people, lawyers included, have ever really tried to establish what laws do exist and how they can be used. It is not enough to complain that the Environment Conservation Act of 1989 does not go far enough. One needs to examine more closely the details of South African environmental law to understand its shortcomings and possibilities.

The environmental laws fall into four broad categories: common law, statute law, international law and administrative law. Each of these is examined here in turn to see how they can best be made to serve the needs of the environment.

Common law

Interdicts and damages actions are the main remedies in common law. Interdicts are used to stop people from continuing on a certain course of action, whereas damages actions are used to obtain compensation for harm caused by that course of action. The common law remedies date back to Roman times and are established in our law.

Unfortunately, there are few examples of reported cases where lawyers have made use of these remedies to protect the environment, and most of these cases relate to noise pollution. But the principles established in these cases can be used to address most aspects of environmental abuse. The main principle emerging from the noise pollution cases and confirmed by the Appellate Division of the Supreme Court in the 1963 case of Regal v African Superslate (Pty) Ltd is that the rights of landowners to do what they wish on their property are not unlimited. Essentially, the courts will come to the rescue of anyone whose occupation of property is unreasonably disturbed by the activities of a neighbour, be it by noise, smoke, river pollution or any other form of despoliation.

However, there are two major drawbacks associated with the common law remedies.

The first is the exorbitant cost of litigation. The victim is often merely an individual and the polluter is often a multinational company or other powerful party which can afford to drag the case out until the victim is bankrupted.

The second problem is the difficulty of proving one's case. The victim seldom has sufficient access to the scientific expertise necessary to prove that the pollution complained of has in fact been caused by the polluter, that it has harmed the victim and that it can be stopped by reasonable means.

The victim seldom has access to the scientific expertise necessary to prove that the polluter is guilty

These problems are not insurmountable. Nevertheless, where for example, chemicals distributed by a multinational are suspected of harming the environment, it may prove foolhardy to attempt to challenge the chemical corporation because of the company's massive resources. In a recent Natal case, the Natal Fresh Produce Growers' Association (a group of vegetable farmers in the Tala Valley) found this to their cost when they unsuccessfully tried to stop seventeen chemical companies from producing and distributing hormonal herbicides in South Africa because of the damage they allegedly cause to broad-leafed crops. The Supreme Court was not even prepared to hear the case as it found the farmers' claims — that any use of the herbicides anywhere in the country would damage fresh produce growing in the Tala Valley — to be 'based on fantasy rather than fact'.

A different and potentially less hazardous route involves instituting interdict proceedings actions against individual users of the chemicals, who are likely to substitute them for less harmful alternatives rather than face expensive litigation. As a general rule, interdict proceedings have more chance of success than damages actions, because the person being interdicted has less to lose than does the person who is being sued for damages.

A new use for the common law remedy of an interdict was discovered a few years ago by lawyers acting for a group of nomadic small livestock farmers in the Richtersveld in northern Namaqualand (see Chapter 8). These pastoralists had been managing a vast area of the Richtersveld for many years by moving their animals according to the seasons, before their grazing could do any long-term damage to the environment. But the National Parks Board had plans to fence a portion of the traditional grazing lands and establish a national park. The pastoralists' attorneys brought an urgent application to the Supreme Court to stop the fencing. The matter was never argued in court, so no mention of the case will be found in the *South African Law Reports*. But the application forced the National Parks Board to the negotiating table, culminating in plans to establish a far larger national park than originally contemplated under the joint management of the National Parks Board and representatives of the local population.

Statutes and regulations

South Africa has a vast and confusing array of statutes and regulations affecting the environment, directly or indirectly. The most notable of these are the Conservation of Agricultural Resources Act, the Atmospheric Pollution Prevention Act, the Water Act, the Environment Conservation Act and the regulations published under these statutes. These are laws and regulations aimed at a wide variety of forms of land abuse and pollution.

It is impossible in this short chapter to examine these laws in detail. However, some general observations are necessary. The four acts mentioned above are administered by four independent departments of government, with the result that the public seldom knows where to report incidents of environmental abuse; and even if such incidents are reported, there is little effective action taken against the offenders.

The policing of the air pollution law is a classic example. Most of the worst polluters are defined by regulation as 'scheduled processes' and are exempt from control by local authorities. Yet there are only seven Department of Health inspectors in the entire country to monitor air pollution by the scheduled industries. Industries that are not scheduled fall under municipal control. But few municipalities have the monitoring equipment necessary to prove that an offence has been committed. It is no wonder that a pall of smoke hangs constantly over most of our major cities.

The Environment Conservation Act of 1989, however, constitutes a major step forward, despite some apparent weaknesses. The Act gives the Minister of the Environment wide powers to regulate development. Most important of all, it empowers the Minister to prevent environmental abuse before it starts. In terms of Section 21, the Minister may identify those activities which in his opinion may have a substantial detrimental effect on the environment. In terms of Section 22, no person

FOWL NOISE

THE clucking and crowing of nearly 7 000 fowls day and night caused Mrs D so much distress that she had to be given tranquillizers by her doctor to enable her to cope. Ultimately the Supreme Court came to her rescue, holding that her complaint was not unreasonable.

This somewhat unusual case is a classic example of how individuals can use interdict proceedings to protect their environmental rights. The case confirmed the rule that a landowner will not be allowed to cause an unreasonable nuisance to his or her neighbours. The same rule can be used by a land occupier to prevent a neighbour from allowing any form of pollution to move across the boundary fence.

Mrs D's husband bought 3,3 acres of land 2 kilometres from the Natal North Coast village of Umhlali in 1958. Two years later he built a house on the property and the family settled there in lush and tranquil surroundings. In July 1964 the family living on the plot next door — a plot roughly twice the size of theirs — started a business producing chickens for sale. The clucking and crowing began to irritate Mrs D; so did the smell. But worst of all, their neighbours turned the lights on every morning at about two o'clock in an attempt to increase production.

In April 1965 Mr D complained to the neighbours that their business was seriously interfering with his family's peaceful enjoyment of their property. This was to no avail, so he went to court asking for an order 'directing the defendant to abate the nuisance' and an interdict 'restraining the defendant from continuing the said nuisance.'

The court found in favour of Mr D. It ordered the chicken breeders to close their existing breeding pens and cockerel pens. In coming to its decision, the court applied the test of the 'reasonable man' (sic). It confirmed that one could be guided by the reactions of the 'reasonable man' who, according to ordinary standards of comfort and convenience and without any peculiar sensitivity to that particular noise, would find it, if not quite intolerable, a serious impediment to the ordinary and reasonable enjoyment of his property. ❐

De Charmoy versus Day Star Hatchery (Pty) Ltd, 1976 (4). South African Law Report, *Durban and Coast Local Division, p 188*

shall undertake an activity defined in Section 21 without written authorization from the Minister or his delegated official. Such authorization may be issued only after consideration of an environmental impact study. The penalty for carrying on such activity without permission is a maximum fine of R100 000 and/or ten years in prison. Regrettably, the Minister has not to date identified any such hazardous activity. The end result is that numerous hazardous industrial enterprises are being established daily without any real checks.

Green activists could lobby the Minister to identify hazardous activities, and in particular those areas of industrial development that are potentially the most harmful to the environment.

Section 2 of the Act allows the Minister to publish environmental policy which could, if utilized, effectively have an impact. Section 3 of the Act reads: 'Each Minister, Administrator, local authority and government institution upon which any power has been conferred or to which any duty has been assigned in connection with the environment by or under any law, shall exercise such power and perform such duty in accordance with the policy referred to in Section 2.' Section 40 further states that the provisions of the Act shall bind the state, including any provincial administration. These provisions read together mean that any decision by any arm of government that affects the environment negatively may be challenged in the Supreme Court as unlawful if it contradicts the published environmental policy.

The rights of landowners to do what they wish on their property are not unlimited

A shortcoming of the Act is that it does not address the problem of *locus standi*. In terms of common law, only a person who can prove a special interest beyond that of others has standing to institute legal proceedings. (More is said about this under Administrative Law.)

Another shortcoming of the Act is that it is to a large extent simply an outline of what can be done once the Minister has published his regulations. Until the regulations are published, the effect of the Act will be minimal. Noise pollution regulations have been published, but these are binding only on cities or towns that have chosen to abide by them. Regulations regarding waste management and environment impact reports had still not been published at the time of writing. This is another area where activists need to press for action by the Minister.

People seeking to establish whether the law covers a specific area of environmental abuse should do one or more of the following:

- Telephone the local municipal health department.
- Consult an attorney.
- Contact an environmental action group.
- Examine the relevant Act of Parliament, if known.
- Consult the chapter on environmental law in *The Law of South Africa*, Volume 9 (Butterworths, 1979), or *Environmental Concerns in South Africa* by RF Fuggle and MA Rabie (Juta, 1983).
- Contact the media.

The victim is often an individual. The polluter is often a multinational corporation

International law

International law is that branch of law which governs relations between states. It is established by agreement between states. The prime object of international law is to prevent conflict by upholding the sovereignty of states. However, intervention has at times been deemed necessary in cases of gross violation of human rights, hence the imposition of sanctions on South Africa by the international community in response to the policy of apartheid. This may well be one of the reasons why South Africa has shown scant respect for international law in the past. This tendency seems to be changing.

There is no effective mechanism for policing international law. States choose to adhere to international law because of the advantages of international co-operation. However, in practice, countries tend to regard the decisions of the International Court of Justice as binding upon themselves despite the absence of means of enforcing such decisions.

The preservation of the environment has not been a major concern of international law for very long, but in recent years numerous international conferences and conventions have attempted to create agreement between states on international environmental law, specifically marine pollution control, ocean-dumping, sea-bed exploitation, trans-frontier air pollution, weapons and weapons testing, the hunting of whales and seals and the protection of Antarctica. The more countries that reach agreement and the greater the number of issues agreed upon, the greater will be the prospects of legal action against offending states.

International environmental law has achieved some spectacular results, the most notable of which was the successful appeal by Australia and New Zealand to the International Court of Justice to stop France from conducting atmospheric nuclear tests in French Polynesia and the Pacific in 1974.

A new and exciting area of international law seems to be developing. Multinational corporations face possible prosecution in their home countries for environmental abuse elsewhere. The area in which the law is likely to develop most rapidly in this regard is that of the international waste trade. (See Chapter 11.) But the development of the law will depend very largely on the flow of information from the waste-receiving states to environmental lobbyists in the waste-producing states. South Africans may not be able to prosecute waste-importers here, but they may be able to facilitate the prosecution of waste-exporters elsewhere by transmission of vital information abroad.

Administrative law

Most decisions affecting the environment in South Africa are made not by Parliament but by government officials and administrative bodies. Administrative law is that branch of law which controls the decisions or actions of such persons or bodies. Where a decision or action by a government official or administrative body unreasonably affects the rights of an individual, such a decision may be taken on review in the Supreme Court. A major drawback is that one may not simply claim that the environment is being unreasonably harmed by the decision and expect to be heard in court. One must prove *locus standi*, or standing to sue. The applicant must prove a direct personal interest in the issue before he or she will be given access to the court. Thus, for example, an individual would find extreme difficulty in challenging in court a decision to legalize the use of DDT in South Africa, as that individual's interest is no greater than that of any other individual. It is of crucial importance that legislation be introduced to remove the *locus standi* requirement for environmental cases.

Review proceedings are distinct from appeals against a decision of a court of law. The issue is not what decision the official or administrative body should have made but whether the proper procedures were applied when coming to that decision. If the official or administrative body has not acted in accordance with proper procedures, the court will declare the decision unlawful, but the court will not impose its own decision. A better way of ensuring that officials make environmentally sound decisions is to demand access to the decision making process. Interested parties should insist that their opinions, evidence and alternatives be heard before decisions are made affecting the environment. If decisions are then made without taking these into consideration, a court will be far more likely to find the decisions irregular than if no alternatives were timeously presented to the decision makers.

PARK PRESERVED BY MP'S ACTION

A MEMBER of Parliament successfully halted the destruction of Groote Schuur Estate at Rondebosch in the Cape in 1981 by bringing an urgent application to the Supreme Court. The Government was in the process of building a number of official residences on the estate when the local MP, Brian Bamford, objected. Groote Schuur had formed part of the deceased estate of the former Governor of the Cape, Cecil John Rhodes. In his will, Rhodes had provided that no suburban residences could be erected on the property and that public access to the park should be preserved for ever. Bamford's purpose in challenging the State was to protect the park for present and future generations. But, for practical legal reasons, he claimed his personal right of access to the park was being infringed.

Bamford's major hurdle was proving that he had standing to sue. He was faced with a rule set by the Appellate Division in the 1933 case of Roodepoort-Maraisburg Town Council versus Eastern Properties (Pty) Ltd. This read: 'It is incumbent on the party complaining to allege and prove that the doing of the act prohibited has caused him some special damage — some peculiar injury beyond that which he may be supposed to sustain in common with the rest of the Queen's subjects by the infringement of the law.' In his application, Bamford did not even allege that he used the park or intended to use it. This was argued by the advocate acting for the State to be a fatal defect.

However, Bamford was successful, despite the odds. The court distinguished between acts which interfered with a right and acts which were prohibited by law. Because Bamford's right of access was being infringed, he did not have to prove any 'special interest'. This case shows clearly the difficulty a person may face in proving standing to sue the State on environmental matters, but it shows equally clearly that the problem may be overcome if a right can be proved. ❒

Bamford versus Minister of Community Development and Auxiliary Services, 1981 (3). South African Law Report, *Cape Provincial Division, p 1054*

Secrecy enshrined in law

Secrecy, censorship and the withholding of information have been a feature of South African legislation, and have unfortunately extended to environmental law. This is reflected, for example, in the Protection of Information Act 84 of 1982, the Nuclear Energy

Act 92 of 1982 and in the Hazardous Substances Act 15 of 1973. These allow for the withholding of information from the public. This is also reflected in the attitude of state officials who refuse to divulge details of permits granted under the Mines and Works Act 27 of 1956 and the Mining Rights Act 27 of 1967. Such refusal meant recently that the inhabitants of the village of Ashburton near Pietermaritzburg were unable to take action against a local quarry because they could not ascertain whether the quarry in question was acting in accordance with the provisions of its permit or not. Until such time as we have legislation similar to the United States Freedom of Information Act of 1966, private individuals will continue to experience serious difficulty in obtaining the information that is vital for the protection of environmental rights.

Access to information is as important to environmentalists as it is to political activists. These groups should work together to change the law in this regard. (See Chapter 1.)

Constitutional law

Constitutional protection for the environment in South Africa is fast becoming a real possibility. Already a number of European countries, India, some American states and more recently Namibia have incorporated environmental protection into their constitutions.

Article 95 of the Namibian constitution provides that the state shall adopt policies aimed at

> maintenance of ecosystems, essential ecological processes and biological diversity of Namibia and utilization of living natural resources on a sustainable basis for the benefit of all Namibians, both present and future: in particular the government shall provide measures against the dumping or recycling of foreign nuclear and toxic waste on Namibian territory.

It is likely that any law passed by the Namibian legislature that comes into conflict with this principle could be declared unconstitutional by the judiciary and it is certain that any executive action by the Namibian government that goes against this principle can be overturned on review by the courts.

Few municipalities have the necessary monitoring equipment. It is no wonder that a pall of smoke hangs over our major cities

Under the present constitution of South Africa, no act of parliament may be challenged in court as illegal. However the courts will review decisions of administrative bodies which interfere unreasonably with the rights of individuals. There is nothing to prevent parliament from immediately defining environmental rights by an amendment to the Environmental Conservation Act. Likewise there is nothing to prevent parliament from legislating that the provisions of the Environmental Conservation Act should prevail over any other laws in the event of a conflict of laws. By making these simple legislative changes, parliament could ensure that in any dispute involving the environment the courts would be obliged to lean heavily in favour of environmental protection.

In the long term, when a new constitution is drafted for South Africa, environmental rights could be enshrined in various ways. The preamble could establish the importance of the environment, and there could be a clause devoted to the environment in the section devoted to the aims of the state (such as in the Namibian constitution). An environmental code could be outlined in a bill of rights and an environmental ombudsman could be appointed, whose function would be to act as a watchdog over environmental abuse by the state. Reference in the constitution to the importance of the environment would mean that in interpreting statutes the courts would be bound to apply an interpretation that provided for environmental protection to any legislation.

Multinational corporations now face possible prosecution in their home countries for environmental abuses elsewhere

Some conclusions

In summary, there is much that can be done within the existing legal framework. The public, and lawyers in particular, need to become more aware of the legal options available and to become more creative in applying these to the widest possible range of environmental abuses. At the same time the responsible authorities need to acquire more sophisticated monitoring equipment and employ more people with scientific expertise who are capable of policing the existing laws.

SELECTED REFERENCES

Chapter 1

R Bahro, *From Red to Green* (Verso, 1984)

F Capra and C Spretnak, *Green Politics* (Hutchinson, 1984)

G Coleman, 'The Campaign Against Thor Chemicals: Trade Unions and the Environment', pp 67-75, in *Critical Health*, no 33, November 1990

F Khan, 'Involvement of the Masses in Environmental Politics', pp 36-38, in *Veld and Flora*, vol 76 (2), June 1990

M Gandar, Earthlife National Congress (Johannesburg: June 1990)

J Glazewski, 'A New Environmental Conservation Act: An Awakening of Environmental Law?' in *De Rebus*, November 1989

S Hazarika, *Bhopal: The Lessons of a Tragedy* (Penguin, 1987)

P Hynes, *The Recurring Silent Spring* (Pergamon Press, 1989)

E Papadakis, *The Green Movement in West Germany* (Croom Helm, 1984)

J Porritt, *Seeing Green* (Hutchinson, 1984)

A Sachs, 'Conservation and Third Generation Rights: The Right to Beauty', in *Protecting Human Rights in a New South Africa* (Oxford, 1990)

K Sale, *Resurgence*, September/October 1989

P Semark, 'Prerequites for Successful Nuclear Generation in Southern Africa', pp 335-337, in *The South African Mechanical Engineer*, vol 40, no 8, August 1990

L Timberlake and L Thomas, *When the Bough Breaks ... Our Children, Our Environment* (Earthscan Publications, 1990)

M Tobias (ed), *Deep Ecology* (Avant Books, 1988)

Chapter 3

Atmospheric Pollution Prevention Act, no 45 of 1965, part IV

J Bourbeau, P Ernst, J Chrome, B Armstrong, M R Becklake, 'The Relationship Between Respiratory Impairment and Asbestos-Related Pleural Abnormality in an Active Work Force', *American Review of Respiratory Diseases*, 1990, pp 837-842

M A Felix, Z M Mabiletja, L Roodt, Unpublished report

M A Felix, Z M Mabiletja, L Roodt, A W J Carlin, M Steinberg, 'The Aftermath of Asbestos Mining - Health Effects of Fibres in the Environment', Proceedings of the first IUAPPA Regional Conference on Air Pollution, vol 2, paper 81

G Reid, D Kielkowski, S D Steyn, K Botha, 'Mortality of an Asbestos-Exposed Birth Cohort' in *South African Medical Journal*, 1990, no 78, pp 584-565

R E G Rendall, M A Felix, E Mogomotsi, Z M Mabiletja, 'Asbestos in the Air Around Abandoned Mines in the North-Eastern Transvaal', Proceedings of the first IUAPPA Regional Conference on Air Pollution, vol 2, paper 80

A B Zwi, G Reid, S P Landau, D Kielkowski, F Sitas, M R Becklake, 'Mesothelioma in South Africa, 1976-84, Incidence and Case Characteristics' in *International Journal of Epidemiology* 1989, no 18, pp 320-329

Chapter 4

E Buch, 'Health Hazards in Botleng'

E Buch and J Doherty, 'Environmental Health Services in Townships: Some Lessons from Soweto' (Centre for the Study of Health Policy, Department of Community Health, University of the Witwatersrand)

C Dingley, 'Electrification as a Solution to Residential Air Pollution', (Department of Chemical Engineering, University of Cape Town. Unpublished paper)

LJ Heyl, 'Smoke Pollution in Urban Residential Areas in South Africa and Possible Solutions',

Proceedings of the International Conference on Air Pollution, November 1988
B Huntley, R Siegfried and C Sunter, *South African Environments into the Twenty-First Century* (Human & Rousseau/Tafelberg 1989)
National Land Committee, 'Land Update No 3', August 1990
E Kemeny, RH Ellerbeck and AB Briggs, 'An Assessment of Smoke Pollution in Soweto', Proceedings of the International Conference on Air Pollution, November 1988
D Simpson, 'Water Pollution' in *Rotating the Cube* (Indicator SA, 1990)
T Wilson, 'Health Services Within Soweto', Carnegie Conference Paper no 170
Y von Schirnding, 'The Effect of Air Pollution on Health: The Ten Year Birth Cohort Study', Paper 10, World Environment Day Workshop, Funda Centre, Soweto, 7 June 1990

Chapter 5

M Berer, 'Evolution of Population Politics' in M Berer (ed), *Women's Global Network on Reproductive Rights: Fifth International Women and Health Meeting: Special Edition*, (Amsterdam, 1987)
B Campbell, 'Indebtedness in Africa: Consequence, Cause or Symptom of the Crisis?' in B Onimode (ed), *The IMF, The World Bank and African Debt Vol 2: The Social and Political Impact* (Institute for African Alternatives and Zed Books, 1989)
J Cock, E Emdon and B Klugman, 'Child Care and the Working Mother: A Sociological Investigation of a Sample of Working Women' (Report submitted to the Carnegie Conference on Poverty, Cape Town, 1983)
J Collinge, 'Population Growth: State Control or Free Love?' in *The Weekly Mail*, 5 - 11 October 1990
C Delph, 'Fewer Children is the Only Way: Reader's View' in *The Star*, 7 September 1990
Department of National Health and Population Development, *The Population Development Programme* (1987)
R Dumont and N Cohen, *The Growth of Hunger: A New Politics of Agriculture* (Marion Boyars, 1980)
Endangered Wildlife, September 1990
S George and N Paige, *Food for Beginners* (Writers' and Readers' Publishing Cooperative Society, 1982)
B Hartmann, *Reproductive Rights and Wrongs: The Global Politics of Population Control and Contraceptive Choice* (Harper and Row, 1987)
D Harvey, 'Population, Resources and the Ideology of Science' in *Economic Geography*, no 50, 1974
B Huntley, R Siegfried, and C Sunter, *South African Environments into the Twenty-First Century* (Human and Rousseau/Tafelberg, 1989)
J Keenan and M Sarakinsky, 'Black Poverty in South Africa', *South African Labour Bulletin*, vol 12, no 3, 1987
B Klugman, *Decision-Making on Contraception Amongst a Sample of Urban African Working Women* (M A Dissertation, University of the Witwatersrand, 1988)
A McKenzie, *The Relevance of the Demographic Transition Theory to South Africa* (Degree of Masters in Public Health, Leeds, 1988)
A McLaren, *Reproductive Rituals* (Methuen, 1984)
I Niehaus, 'Relocation into Phuthaditjhaba and Tseki - A Comparative Ethnography of Planned and Unplanned Removals' in *African Studies*, vol 48 no 2, 1989
R Omond, *The Apartheid Handbook* (Penguin, 1985)
M Redclift, *Development and the Environmental Crisis* (Methuen, 1984)
F Sai, 'Population and Health: Africa's Most Basic Resource and Development Problem', in R Berg and J Whitaker (eds), *Strategies for African Development* (University of California Press, 1986)
R Scholes, *Global Change and its Implications for Food Security in Southern Africa* (Paper presented at the AIESEC Conference, Johannesburg, 1989)
Science Committee of the President's Council, *Report on Demographic Trends in South Africa*, 1983
United Nations, 'World Population Plan of Action' in *The Population Debate: Dimensions and Perspectives: Papers of the World Population Conference, Bucharest*, vol 1 1974/1975
F Wilson and M Ramphele, *Uprooting Poverty: The South African Challenge* (David Philip, 1989)

Chapter 6

African National Congress, 'Future Environmental Policy for a Changing South Africa' (Discussion paper prepared by M Sisulu and S Sangweni, November 1990)
D Cowen, 'Towards a Philosophy of Law Adequate to Safeguard the Environment' (Unpublished article, 1990)
R Crompton, 'Trade Unions as Environmental Watchdogs' (Unpublished article, 1990)

DGB, 'Environmental Protection and Qualitative Growth: Combating Unemployment and Speeding up Qualitative Growth Through Increased Environmental Protection' (Translated from German into English by the European Trade Union Institute, Brussels, 1985)

Financial Mail, SASOL Survey, 21 September 1990

Fortune, 21 February 1990, pp 24-30

S Gelb, 'Democratizing Economic Growth: Alternative Growth Models for the Future' in *Transformation 12*

S Gelb (ed) *South Africa's Economic Crisis* (David Philip, 1991)

Global Warming Watch (Public Health Institute, New York, July 1990)

Greenpeace, *The Gods Must Be Crazy: Mercury Wastes Dumped by Thor in South Africa* (A Greenpeace International Waste Trade Profile, 1990)

R Herding, 'Greening the Workplace', *International Labour Reports*, no 39, 1990

B Huntley, R Siegfried and C Sunter, *South African Environments into the Twenty-First Century* (Human and Rousseau/Tafelberg, 1989)

ICEF, 'The Role of Workers in Preventing Occupational Hazards', 1990

ICEF, 'Industrial Change and the Environment', World Conference on Occupational Health and the Environment, 1990

International Labour Organization, 'Draft Code of Practice on the Prevention of Major Accident Hazards', January 1990

International Labour Organization, *Encyclopedia of Occupational Health and Safety*, Geneva, 1985

Greenpeace International, 'International Trade in Toxic Wastes: Policy and Data Analysis' in *Toxic Terror: Dumping of Hazardous Wastes in the Third World* (Third World Network, Malaysia)

M Jacobs, 'Green Dilemma' in *New Socialist* no 62, 1989

O P Kharbanda and E A Stallworthy, *Safety in the Chemical Industry: Lessons From Major Disasters* (Heinemann)

D H Meadows, 'Preserving the Earth's Most Valuable Library', *Weekly Mail*, 21 to 27 September 1990

J Meyers and M Steinberg, 'Health and Safety' in *South African Labour Bulletin*, 8(8) and 9(1), 1983

MSF, 'MSF and the Environment: a Policy Strategy Document'(Management Science Finance)

Pan Africanist Congress, 'An Environmental Policy for the Pan Africanist Congress of Azania' (Discussion document by B Desai, November 1990)

A Sachs, 'Conservation and Third Generation Rights: the Right to Beauty' in *Protecting Human Rights in a New South Africa* (Oxford, 1990)

The Sunday Star, 4 November 1990

The New York Times, 14 October 1990

The 1990 Ceres Guide to the Valdez Principles, Boston

Time, The Battle to Save the Planet, 23 April 1990

United Steelworkers of America, 'Report of the Task Force on the Environment', August 27-31 1990

United States Environmental Protection Agency, 'Reducing Risk: Setting Priorities and Strategies for Environmental Protection', 1990

Chapter 8

F Archer, 'Planning With People: Ethnobotany and African Uses of Plants in Namaqualand, South Africa', pp 959-972 (Mitt Inst Allg Bot Hamburg,1990)

F Archer, M T Hoffman and J E Danckwerts, 'How Economic are the Farming Units of Leliefontein, Namaqualand?', *Journal of the Grassland Society of Southern Africa* 6, 4 (December 1989), pp 211-215

The Argus

Atomic Energy Corporation, 'Vaalputs: National Radioactive Waste Disposal Facility'

M Baumann, 'Smokey's store', *Leadership* 8, 9 (November 1989), pp 72ff

S Balic, 'Richtersvelders Win Land Rights Tug-o'-War', *New Ground*, 1 (September 1990), pp 6-7

E A Boonzaaier, 'Economic Differentiation and Racism in Namaqualand: A Case Study', Second Carnegie Inquiry Conference Paper 68 (SALDRU, 1984)

E A Boonzaaier, M T Hoffman, F Archer and A B Smith, 'Communal Land Use and the "Tragedy of the Commons": Some Problems and Development Perspectives with Specific Reference to Semi-Arid Regions of Southern Africa', *Journal of the Grassland Society of Southern Africa* 7, 2 (June 1990), pp 77-80

J Dunne, 'Towards a Regional Development Strategy for Namaqualand', SALDRU Working Paper 75 (SALDRU, October 1988)

M L Borchers, F Archer, A Eberhard, 'Namaqualand Household Energy Survey: A Study of

Energy Consumption Patterns and Supply Alternatives in Six Reserves' (Centre for Research into Appropriate Energy Technologies, University of Cape Town Energy Research Institute, October 1990)

'Nuclear Power Station Sites: Purchasing of Sites, Southern Cape', (ESKOM, June 1990)

'Nuclear waste: A Problem Solved', (ESKOM, December 1985)

'Far Away, But Wide Awake', *Learn and Teach* 3 (1990), pp 22-27

R Hill, 'Statement on Ministers' Council Decision Not to Sign the Contract for the Richtersveld National Park', Cape Town, 9 November 1990. Reproduced in a publication of the SPP/Richtersveld Gemeenskap Komitee, 1990

R Hill, F Archer and L Webley, 'Integrated Environmental Management in the Conflict Over Change in Land Tenure in the Reserves of Namaqualand, South Africa', Proceedings of the VII Annual Meeting of the International Association of Impact Assessment, Brisbane, 5-9 July 1988 (In press)

'Klagtes in Verband met die Wegdoening van Radioaktiewe Afval te Vaalputs', Deposition of communities in Leliefontein reserve, Klipfontein, 1990 (Unpublished)

E Koch, D Cooper and H Coetzee, *Water, Waste and Wildlife: The Politics of Ecology in South Africa* (Penguin, 1990)

Koeberg Alert, 'Vaalputs Radioactive Waste Disposal Facility', Koeberg Alert Information Sheet, Cape Town, 1989 (Unpublished)

Koebert Alert Research Group, 'The Power of the State and the State of Power: Recent Developments in South Africa's Nuclear Industry', pp 469-494, in Moss and Obery (eds), *South African Review 4* (Ravan, 1987)

H Kröhne, 'A Rural Development Strategy for an Arid Region: The Case of the Richtersveld', (Unpublished MCRP dissertation, School of Architecture and Planning, University of Cape Town, 1988)

H Kröhne and L Steyn, 'Grondgebruik in Namakwaland' (Surplus Peoples Project, 1990)

D Luyt, 'The Transition to Capitalism in the Namaqualand Reserves (B Soc Sci Hons dissertation, University of Cape Town, 1981. Unpublished)

R Meintjes, 'Richtersveld se Mense Vat 'n Kans', *Die Suid-Afrikaan* no 24 (December 1990), pp 30-33

Namaquanuus

S Posnik, 'Environmental Assessment of the Radioactive Waste Disposal Facility at Vaalputs, Bushmanland', pp 36-41 in CSIR, Proceedings of the South African Institute of Ecologists Symposium on Nuclear Energy and the Environment in South Africa, Pretoria (CSIR, 1 February 1985)

Rapport

'The Scramble for a Nuclear-Ridden Africa', *Fission Chips* no 2, 2 (August 1990), pp 1-4

J Sharp, 'Rural Development Schemes and the Struggle Against Poverty in the Namaqualand Reserves', Second Carnegie Inquiry Conference Paper 69 (SALDRU, 1984)

J Sharp and M West, 'Land, Labour and Mobility in Namaqualand', Second Carnegie Inquiry Conference Paper 71 (SALDRU, 1984)

L Steyn, 'Privatization and the Dispossession of Land in Namaqualand', pp 415-425 in Moss and Obery (eds), *South African Review 5* (Ravan, 1989)

Surplus Peoples Project, Annual Report 1988/9

Surplus Peoples Project, Annual Report 1989/90

Surplus Peoples Project, 'Port Nolloth: Nowhere to Go' (SPP Factsheet 5, June 1988)

Surplus Peoples Project, 'Richtersveld National Park?' (SPP Factsheet 7, June 1989)

Surplus Peoples Project and Richtersveld Gemeenskapskomitee, (Dossier: Richtersveld National Park, 10 November 1990)

C Terreblanche, 'Die Richtersveld', pp 8-9 in *Vrye Weekblad Ekologiebylae*, 16 November 1990

W Thomas, 'Voorgestelde Benadering tot die Besluitneming en Bestuur van die Namakwalandse "Diamantfons"', Cape Town, 10 August 1990 (Unpublished)

A Wildschut and L Steyn, 'If One Can Live, All Must Live: A Report on Past, Present and Alternative Land use in the Mier Rural Reserve in the Northern Cape' (Surplus Peoples' Project, 1990)

Chapter 12

P Blaikie, *The Political Economy of Soil Erosion in Developing Countries* (Longman, 1985)

D Cooper, 'Working the Land', *New Ground* (Environmental and Development Agency, Johannesburg, 1988)

'The "Aggro" Chemicals', *Critical Health* no 33, November 1990

'Homelands: Failure of the Dreaming Mind', Special edition of *Cross Times*, Cape Town, 1990

H Dolny, 'The Development Bank of South Africa(DBSA): A Review of and Commentary on

Their Policy Proposals' in *Mimeo*, November 1990
A Durning, 'Apartheid's Environmental Toll', Worldwatch Paper 95 (Worldwatch Institute, Washington, 1990)
'Pesticides' in *Mimeo* (Greenpeace International, Amsterdam)
'Plan for Huge Transfer of Land', *Sunday Times*, 18 November 1990
B Huntley, R Siegfried and C Sunter, *South African Environments into the Twenty-First Century* (Human & Rousseau/Tafelberg, 1989)
'Environmental Strategies in the 1990s' in *Rotating the Cube* (Indicator SA, 1990)
E Koch, 'Government Tests Reveal Abuse of Agent Orange', *Weekly Mail*, 23-29 March, 1989
N Myers (ed), *Gaia: An Atlas of Planet Management* (Pan, 1985)
L Timberlake, *Africa in Crisis* (Earthscan, London, 1985)

Chapter 16

Chapter 16 has made extensive and liberal use of the information and ideas published by McNeely et al (1990); this source is gratefully acknowledged.
R Boycott, 'A New Frog From the South-Western Cape Province', *African Wildlife* no 44 (6), 1990, pp 343 - 346
WR Branch (ed), *South African Red Data Book - Reptiles and Amphibians*, South African National Scientific Programmes Report 151, 1988
RK Brooke, *South African Red Data Book - Birds*, South African National Scientific Programmes Report 97, 1988
AA Ferrar, 'The Role of Red Data Books in Conserving Biodiversity' in *Biotic Diversity in Southern Africa: Concepts and Conservation*, ed BJ Huntley, pp 136 - 147 (Oxford, 1989)
C Hilton-Taylor and A Le Roux, 'Conservation Status of the Fynbos and Karoo Biomes' in *Biotic Diversity in Southern Africa: Concepts and Conservation*, ed BJ Huntley, pp 202-223 (Oxford, 1989)
AV Hall, M de Winter, B de Winter B and SAM van Oosterhout, 'Threatened Plants of Southern Africa', South African National Scientific Programmes Report no 45, 241 pp
AV Hall and HA Veldhuis, *South African Red Data Book: Plants - Fynbos and Karoo Biomes*, South African National Scientific Programmes Report no 117
AV Hall, 'Rare Plant Surveys and Atlases' in *Biotic Diversity in Southern Africa: Concepts and Conservation*, ed BJ Huntley, pp 148 - 156 (Oxford, 1989)
BJ Huntley (ed), *Biotic Diversity in Southern Africa: Concepts and Conservation* (Oxford, 1989)
B Huntley, R Siegfried and C Sunter, *South African Environments into the Twenty-First Century* (Human & Rousseau/Tafelberg, 1989)
F Kahn, 'Beyond the White Rhino: Confronting the South African Land Question', *African Wildlife* no 44 (6), 1990a, pp 321-324
F Kahn, 'Involvement of the Masses in Environmental Politics', *Veld and Flora*, June 1990b, pp 36-38
IAW Macdonald, 'Man's Role in Changing the Face of Southern Africa' in *Biotic Diversity in Southern Africa: Concepts and Conservation*, ed BJ Huntley, pp 51-77 (Oxford, 1989)
JA McNeely, KR Miller, WV Reid, RA Mittermeier and TB Werner, *Conserving the World's Biological Diversity* (International Union for the Conservation of Nature and Natural Resources, World Resources Institute, Conservation International, World Wildlife Fund-US, and the World Bank, 1990)
S Schneier, 'Konflik Oor die Kus', *Conserva* no 5 (6), 1990, pp 12-13
WR Siegfried, 'Preservation of Species in Southern African Nature Reserves' in *Biotic Diversity in Southern Africa: Concepts and Conservation*, ed BJ Huntley, pp 186 - 201 (Oxford, 1989)
PH Skelton, *South African Red Data Book - Fishes*, South African National Scientific Programmes Report no 137, 1987
RHN Smithers, *South African Red Data Book - Terrestrial Mammals*, South African National Scientific Programmes Report 125, 1986

INDEX

acid rain 66, 98, 132, 135, 140, 143, 146-8
administrative law 248-9
African Explosives and Chemical Industries (AECI) 60, 134-5, 151, 164
African National Congress 12-16, 128
African Skimmer 239
agriculture 176-92; indigenous 181-2, 188
air pollution 4, 13, 56-7, 98, 139-57, 246
Albertyn, Chris 18-19, 28
Alexander Bay 24
Alexandra township 48-53
anti-nuclear lobby 125
Anti-Pollution Appeal Committee 159
apartheid 12-13, 17, 64, 72-7, 84, 196-7, 239, 248
asbestos 21, 33-43
Association for Rural Advancement 223-7
atmospheric pollution *see* air pollution
Atmospheric Pollution Prevention Act (1965) 40, 246
Atomic Energy Corporation 30, 100-1, 121, 123, 124, 125
auxiliary game guard system 220
Azaadville 24, 233
Azanian People's Organization 31-2

bacterial contamination of water 137
Barberton 135-6
Basel Convention 84, 124, 168
beach apartheid 196-7
Bill of Rights 15-16, 65, 251
biodiversity 230-41
biological resources 233
birds 238
birth control 67, 74, 77
blue swallow 238
Bophuthatswana 177
Boraine, Andrew 49-50
Botleng 54, 57
Buthelezi, Mangosuthu 225

cancer 3, 102, 104, 141-2, 154
Cape floristic kingdom 234, 235, 236
Cape Inventory of Critical Environmental Components 236
capitalism, 7, 182
carbon dioxide pollution 99-100, 146, 154-5
Carson, Rachel 5
Cato Ridge 82-3, 169-71
chemical weapons 9
Chemical Workers' Industrial Union 10, 25-6, 83
Chernobyl 6, 78, 102
chlorination of water 137
chlorofluorocarbons 151-4
Ciskei 171, 177, 182
civic organizations 62
civil society and collective rights 89-91
climatic effects of pollution 99-100
coal 97-100, 143-4, 145-50
  domestic use 4, 51, 56, 148, 150
coastal development 197-8
coastal zone 194-9
common law 244-5
communal areas' economy 190
communal land ownership 118
communal lifestyles and conservation 23
community
  consultation/involvement 10-12, 20-32, 83, 213-14, 222, 224, 227
  development 62
  empowerment 9-12, 222
  environmental rights 82-4
  opposition to reserves 210-13
  participation in conservation 14, 23-4, 119-20, 213-22, 226
compensation for removals 224, 225-6
Congress of SA Trade Unions 10, 85-6
Congress of Traditional Leaders of SA 23
Conservation of Agricultural Resources Act 246
constitution and green rights 15
constitutional law 250-1
consumer resistance 90
contraception (birth control) 67, 74, 77
copper 114-15, 143
Crocodile River 182
Crompton, Rod 10, 26
crop spraying 185
crops affected by pollution 98, 148 *bis*
Curry, Minister David 121
cycads 6

dagga 27, 185
Dalton Asbestos Mill 33, 42
DDT 132, 185
deaths, pollution-related 98-9
deep ecology 16
defoliants 27, 185
deforestation 70, 131
degradation of land 13, 131-2, 176-7
Delmas 57
Democratic Party 4
demolitions 55-6

Department of Defence 9
Department of Environment Affairs 9
Department of Planning and Environment 104
Department of Water Affairs 205
De Pontes, Peet 121-4
desulpherization of coal 149
developing countries: waste imports 168-9
Development Bank of SA 178, 187, 188-9
development corporations 62
diamond industry 114-15, 198
Dieldrin 132
Dobsonville squatters 55-6
dolomitic ground 57, 61
Dolphin Action and Protection Group 10, 26, 208-9
dolphins 206, 208
dumps and dumping 4, 61, 132-4, 162, 163
dunes 198
Durban 57, 106

Earthlife Africa 11-12, 14, 18-19, 24, 25, 27, 30, 83, 122, 134, 137, 162
Eastern Transvaal Highveld 140-2, 144-9
eco-pacifist movement 7
Ecology party 4
economic growth rate and environment 84
ecosystems 87, 230, 231, 236
education and birth control 77
Eerste River 204
Egyptian vulture 239
Elands River 134
electricity conservation programme 150
electricity consumption and pollution 53
electricity generation 97-108
electricity supply 51, 105
elephant tusks/ivory 6, 14
elephants 210-11, 230
employment 87-9, 220-1
Endangered Wildlife Trust 218
energy 94-109
    conservation 108
    consumption 101, 104-5
    conversion 80
    efficiency 9, 104, 108
    –environment relationship 80, 95-6
    –growth fallacy 104
    integration 106-8
sources, alternative 13, 106
waste 165 (*see also* nuclear waste)
enforcement of conservation 6, 252
Environment Conservation Acts 5, 15, 196, 199, 244, 246-7, 250
environmental abuse (international law) 248
environmental benefits 87
Environmental Development Agency 44-5
Environmental Film Workshop 10, 14
Environmental Protection Agency 5-6, 29, 158

ESKOM 101, 122-8 *passim*, 145-50 *passim*
estuaries 198-9
Exclusive Economic Zone 194
excreta-related infections 53-4

False Bay 204, 205
fertilizers 132
firewood 94-7, 105, 179, 182
fisheries 199-203, 207
fishes 238
fishing resources, control over 200
fishing zones 194 *bis*
flora 233, 234-5, 236
flue gas 99
Food and Allied Workers' Union 10, 26, 200
food base 233
forced removals 8, 72, 118, 223, 224
forests 98, 131, 148, 183, 186-7, 238
Fourways Gardens Village 46-7, 53
free market 90-91
Friends of the Earth 7
fynbos 236-7

game farming 237
game reserves *see* nature reserves
Gandar, Mark, 110-11
Gazankulu 182
Geach, Bev 174-5
genes 230, 231, 233, 235
gill-netting factor 26, 202, 209
global solidarity 17
global threats to biodiversity 233-4
grassroots campaigning 12, 17, 16, 62
grazing 13, 118, 177, 184
The Green Pages 174
green politics 1, 4, 16
green rights 15-17
Greenbelt Action Group 46
greenhouse effect 100, 154-5
Greenpeace 12, 25
Groote Schuur Estate 249
Gross Domestic Product 104
groundwater contamination 164

Hazardous Substances Act (1973) 250
health 53-4, 75, 77
heat release to atmosphere 146
Hendrickse, Allan 197
herbicides 2, 19, 26, 132, 184, 245
Herero 210-22
Herschel 177
high seas, freedom of 193, 194
Himba 210-22
Holomisa, General Bantu 171
homelands 13, 176-83
    agricultural restructring 189-90

capitalist penetration 182-3
overpopulation 72-3
soil erosion 177-9
toxic waste shipping to 171
House of Representatives 116
household waste 161, 173
housing 50, 51-2
housing trusts 62,
human beings in nature 16-17, 31-2, 223-7, 232, 239-40
human rights and international law 31, 81
hydroelectric power 106, 108, 125, 155

incineration 164-5
industrial democracy 81-91
industrial production mortality rate 82
industrial waste *see* waste
infant mortality 2, 7, 75-6
information, freedom of 29-30, 250-1
Ingwavuma 223
initiation and population control 74
Integrated Pest Management 132-3, 186
interdict proceedings 245
international law 248
International Federation of Chemical, Energy and General Workers' Unions 10
International Union for the Conservation of Nature and Natural Resources 231, 232
intertidal zone 195
ivory/elephant tusks 6, 14

Johannesburg 137, 152, 161
Jukskei River 50, 51
Julies, Dr Andrew 120

KaNgwane 23, 182
Karoo 184, 237, 238
Katlehong 163, 173
Khayelitsha 67
Khumbane, Tshepo 44-5
Koeberg Alert 122
Koeberg Power Station 6, 122, 124, 126
Komaggas 24, 122-3, 125, 127
Kosi Bay Nature Reserve 23, 224, 225-7
Kotze, Gert 4, 170
Kruger National Park 4
kwashiorkor 179
KwaZulu 179, 181, 183
Kwazulu Bureau for Natural Resources 23, 182, 223, 224

labour and environment 78-91
Labour Party 116-17, 120
land crisis 182
land degradation 13, 131-2, 176-7
land issue 28, 113, 128
land ownership, communal 118
land redistribution 65, 86, 187-9
land trusts 51-2, 62
land use 176-92
landfilling 133, 164
law and environmental protection 6, 244-51
Law of the Sea Convention 193, 194
Lebowa 177
legal procedure 249
Lekgheto, Japhta 64-5
Leliefontein Reserve 116-17, 122, 124
Letaba River 182
Lethaba power station 21
leukaemia 102, 104, 143, 233
Liesbeeck River 133 illus
Lindane 132, 133, 185
linefishing 200
livestock farming 181-2, 188, 191
loan funds, community 62
local goverment 52, 116
local-level negotiations 62
locus standi 249
Lorimer, Rupert 4
lung cancer 141-2, 148
lung conditions, pollution and 37-9, 43, 141

Mabuyakhulu, Mike 32
Machinery and Occupational Safety Act 81
Mafefe 21, 33-43
Makurung 20, 29
malnutrition 115
mammals 234, 237-8
management boards 116-17, 119
Mandela, Nelson 117
Maputaland 22-3
marine environment 193-207
marine pipelines 206
marine pollution *see under* Oceans
marine resources 194, 201-3
market forces 90
Mass Democratic Movement 197
mass environmental awareness 14, 16
Matjieskloof 10
Mdakane, Richard 52, 53
measles 54
Media Training and Development Fund 127
Medical Research Council 148
medicines 233
mercury 3, 25, 28, 82-3, 134-5, 136, 137, 169, 170, 193
mesothelioma 33, 35, 38, 43
methane gas 164
metropolitan local government 62
Mier reserve 117-18
migrant labour 74, 126
military expenditure 8-9
mining 15, 97, 114, 115, 133, 134, 250
Mlangeni, Andrew 12
Mondi 186, 187

mortality 7, 74-6, 98-9
motor vehicle pollution 148, 156-7
Mthethomushwa 23
Mtwalume estuary 199
Mufford, Jenny 158-9
mussels 206

Nama people 113-14
Namaqualand 24, 30, 112-28, 114 map
Namaqualand Council of Churches 127
Namaqualandse Burgervereniging 24, 31 illus, 123, 127
Namaquanuus 30, 123, 127
Namibia constitution 250
Natal Fresh Produce Growers' Association 26
National Association for Clean Air 158-9
National Atlas of Critical Environmental Components 236
National Centre for Occupational Health 37
National Council of Trade Unions 10, 22
National Environmental Awareness Campaign 12, 47, 64
national parks 235
National Parks Board 118-21, 245
National Union of Metal Workers of SA 10
National Union of Mineworkers 8, 122, 126
nationalization of marine resources 203
nature reserves 235, 241
  community and 2, 10, 14, 23-4, 119-20, 210-22, 224-7
Ndaba, Humphrey 92-3
nerve gases 9
nitrogen oxides pollution 98-9
noise pollution 244, 246, 247
Norweto 46-7
Nourivier 122-3, 124
nuclear accidents 102; *see also* Chernobyl
Nuclear Energy Act 29, 250
nuclear power 6, 9, 100-2, 125-6, 155
  ANC policy 14
nuclear tests 248
nuclear waste 6, 24, 30, 102, 121, 124-5
nuclear weapons 6, 8, 9, 14
nutritional diseases 75, 179

Occupational Diseases in Mines and Works Act 35
occupational health and safety 80-1
oceans 193-209
oil-from-coal 100; *see also* Sasol
oil spills and discharges 207
ombudsman 251
Operation Clean Up 64
Operation Hunger 115
organic farming 190-2
organic waste fertilizer 206
organically-grown produce 3
Organization of African Unity 172
overcrowding 178; diseases of 54, 55
overpopulation *see* population
Overvaal Resort 2
Owen-Smith, Garth 214, 217, 218
oysters 206
ozone 151-4

Pan Africanist Congress 14
paraquet 27, 30, 185
parks 13, 52
PCB 206
Peacock Bay 121, 124
pellagra 179
people perspective on environment 2-4
people's park 13
pest management 132-3, 186
pesticides 2, 9, 30, 132, 184-6, 191, 205-6
petrol pollution 156
Phalaborwa 143
Phola Park 106
Pietermaritzburg waste dumping 162-3
pilchard exploitation 201-3
Player, Ian 228-9
plutonium 6, 100
pneumonia 54
political perspective on environment 1, 4
population 1, 66-77; *see also* overcrowding
Port Nolloth 118, 125, 127, 172
poverty 75, 179
power (energy) *see* energy; nuclear power
power station pollution 98-9
privatization 116-18
process modification 166
product substitution 166
progressive developmental perspective 4
Protection of Information Act 250
pulp waste exploitation 206
Purros 10, 23-4, 31, 210-22

QwaQwa 72, 178, 179
quality of life, township, 61

racial bias in conservation 2
radiation 102
radioactive waste *see* nuclear waste
radioactivity 101-2
recycling 52, 108
Red Data Books 231, 235
redistribution of land 65, 86, 187-9
refuse removal 50, 52, 55, 173
refuse-related diseases 54
renewable energy 106
reproduction controls, social 74
reptiles 238
Research Institute for Reclamation Ecology 39, 41, 42
Reservation of Separate Amenities Act 195-7
resettlement 2, 223-7
resources

conservation 108
population balance and 77
usage 68, 69, 70-7
wastage 165
respiratory diseases 57, 98, 140, 141, 158
rhino horn 6, 14
rhinoceros 240
Rice, Nan 208-9
Richtersveld 10, 25 illus, 117, 118-21, 181
riverine rabbit 237
rivers 13, 130, 131, 132, 182
road haulage industry 115
roan antelope 237
Robberg Peninsula 198
Robinson Deep landfill site 164
rock lobster 200
Rooi Els 25
rural areas *(see also* homelands)
degradation 131-2, 176-9
development 23, 189-90
electrification 105, 190
state intervention 179-82

SAAV 243
Sachs, Albie 15, 17, 28
Saldanha Bay 200
salinisation of water 133
Sandton City Council 52
Sappi 134, 186, 187, 205
Sasol 82, 89, 100, 135, 136
Sasolburg 21, 60, 141, 158
Schedule Five Parks 120
Schweitzer, Albert 17
seals 30-1
Seashore Act 195, 196, 197
Secunda 82, 146-7
Sentrachem 60
Separate Amenities Act 196
sewage 204-6
sewerage 50, 55
shipping pollution 207
siltation 132
simplification of life 8
Sisulu, Max 12-13
Sisulu, Walter 12, 13
Skeleton Coast Park 210
skills training 52
skin cancer 154
Small Business Development Corporation 115-16
smoke pollution level 56
social engineering, enforced 181
social relations, disruption of 74
social security and birth rate 76
socialism 7, 78-9
Society Against Nuclear Energy 110
Sodwana Bay 225
soil erosion 13, 28, 131-2, 177-9, 190
soil fertility, protecting 190-1
solar energy 13, 106, 155-6,
Sonchem 25
Songimvelo Nature Reserve 23
South African Chemical Workers' Union 10, 22, 92
South African Defence Force 8, 9
South African Rivers Association 27
Southern African Development Coordinating Conference 106-8
Soweto 4, 55, 89, 147-8, 150-1
species, endangered 230, 231
Springbok 114, 127 illus
squatters 3 illus, 106, 118
St Lucia 4, 9, 32, 198, 228
standard of living 75-6
state and environment 4-5
statutes and regulations 246-7
Steinkopf 117
Steyn, Lala 120
subsistence producers 181
suburbs 53-5
sugar farms 207
sulphur pollution 98-9, 143-50 *passim*
Surplus Peoples Project 117-20 *passim*, 128
sustainability 87, 160, 189-90, 232, 234

Taiwanese trawlers 26, 209
Tala Valley 132, 184, 245
tax base, single 62
Taylor, Aileen 242-3
Tembe Elephant National Park 225-7
Tentedorp 118
territorial waters 193, 194 *ter*
Thor Chemicals 3, 10, 19, 25-6, 28-9, 81, 82-3, 169, 170
titanium 9, 32
tourism 211, 216-17, 218, 221, 235
townships 47-63
toxic substances at the workplace 80
toxic waste *see* waste
trade unions 9-10, 19, 83, 85-6
traditional conservation 227
Transkei 178, 179 illus, 181 illus
Transkei Development Corporation 171
transport infrastructure 61
treeplanting campaigns 52, 154-5
trekboere 113-14
trek-netting 205 illus
Treurnicht, Andries 4
tricameral privatization 116
Trihalomethane 137
Tseki 72-3, 181
Tugela River 146

unemployment and overpopulation 76
uranium 100

urban areas environmental crisis 46-65
Urban Foundation 187-8

Vaal catchment area 134
Vaal River 131, 134, 135
Vaalputs dump 113 illus, 122-3, 124-5
Venda 182
Viva Park 58
vivisection 242-3
Vlok, Adriaan 27

war 8-9
waste 3, 10, 13, 25, 53-4, 66, 83, 121-4, 133, 134, 135, 160-75, 279
  assimilation, environmental 87
  exports and imports 2, 14, 81, 124, 167-8, 171, 248, 250
  legislation 166-7, 248-50
  ocean disposal 205-6
  transportation 165, 168-9
water
  consumption, per capita 137
  management policy 206
  pollution 53-4, 57, 132-8
  legislation 138
  resources 129-38
  shortage 53, 55
Water Act 246
water-related illnesses 53
water table, underground 53
wattled crane 238

West Rand 133
wetlands 238
wild dogs 237
Wilderness Leadership School 228-9
wildlife reserves *see* nature reserves
wildlife resources: community sharing 14
Wildlife Society of Southern Africa 30-1
Winterveld Squatter Camp 136
women's position and birth rate 76, 77
work environment 79, 80, 173
workers, environmental rights of 82-4
World Charter for Nature 231-2, 240

yellow billed oxpecker 239

Zamdela 21-2, 60